# Studies in Computational Intelligence     452

**Editor-in-Chief**

Prof. Janusz Kacprzyk
Systems Research Institute
Polish Academy of Sciences
ul. Newelska 6
01-447 Warsaw
Poland
E-mail: kacprzyk@ibspan.waw.pl

For further volumes:
http://www.springer.com/series/7092

Juan D. Velásquez, Vasile Palade,
and Lakhmi C. Jain (Eds.)

# Advanced Techniques in Web Intelligence-2

Web User Browsing Behaviour
and Preference Analysis

 Springer

*Editors*
Dr. Juan D. Velásquez
Dept. Industrial Engineering
School of Engineering & Science
University of Chile Republica
Santiago
Chile

Prof. Dr. Lakhmi C. Jain
School of Electrical and Information
Engineering
University of South Australia
Adelaide, SA
Australia

Dr. Vasile Palade
Oxford University Computing
Laboratory (OUCL)
Oxford
UK

ISSN 1860-949X
e-ISSN 1860-9503
ISBN 978-3-642-43035-0
ISBN 978-3-642-33326-2 (eBook)
DOI 10.1007/978-3-642-33326-2
Springer Heidelberg New York Dordrecht London

Printed on acid-free paper

Springer is part of Springer Science+Business Media (www.springer.com)

# Foreword

This book presents itself as a logical and coherent sequence of tasks, which should be performed when trying to get information and make sense from the behaviour of Web users in nowadays Web based systems, ubiquitous environments, Web 2.0 systems, social networks and, of course, more traditional Web applications. The main purpose is to create new effective real-world Web based systems that personalize the Web user experience.

The book also explores the frontier of Web Intelligence methods, focusing on the theories and algorithms for analyzing Web user behaviour. The result was a comprehensible state-of-the-art about these issues, which was possible thanks to the collaboration of many experts around the world on this subject, who were asked to reflect on the latest trends in Web Intelligence that could help obtaining a better understanding of the behaviour of Web users.

The book is organized in seven self-contained chapters with a common editorial line, with chapters focusing on Web data collection, data processing, behaviour pattern identification as well as possible applications of related techniques for building Web recommendations systems. The writing is easy to follow in general and the coverage is exhaustive, showing the main concepts and topics in the area. The level is appropriate for both advanced undergraduate and graduate courses, and mainly for researchers with interests on the frontier of Web Intelligence techniques and applications related to Web user browsing and preference identification and analysis.

Finally, I would like to congratulate the editors and the authors of the individual chapters for their marvelous job. We believe that the effort will be rewarding to them and I am sure the book will prove very useful to its readers.

Maebashi Institute of Technology, Japan,                    Professor Ning Zhong
July 2012

# Preface

This research volume is a continuation of our previous volume entitled "Advanced Techniques in Web Intelligence -1", published in 2010. The main aim was to update on the state-of-the-art in the field of Web Intelligence and to present successful real-world applications in this area of research.

The present volume focuses on analysing the web user browsing behaviour and preferences in traditional web-based environments, social networks and web 2.0 applications. Recent web intelligence algorithms and techniques are revised and used to get a better comprehension about the web user needs and desires in the wide Web, tackling the big challenge in the development of new web-based applications: how to personalize the web user experience.

This volume is organized as a sequence of chapters centred around the main design steps when trying to personalize the web user experience in web-based applications. With this in mind, Chapter 2 introduces several techniques for collecting and pre-processing web data, ranging from more traditional ones, based on web log registers, to more recent ones, as the positions of the user's eyes on the screen. Chapter 3 shows how the cognitive science can be applied for the analysis of the web user behaviour in a web site. In Chapter 4, web usage mining technics and algorithms are proposed and applied for discovering usage patterns in web applications. How to get valuable information from what the web users write in social networks and web 2.0 environments is the subject of Chapter 5. Chapter 6 presents how the newly collected knowledge about the web user can be used to create web usage based adaptive systems. Finally, Chapter 7 shows how to create and evaluate web recommendation systems, which is a main goal of personalizing the web user experience in the Web.

We are first grateful to the authors who contributed to this book, without whom this book would have not taken place otherwise. We are also grateful to the reviewers who helped us during the preparation of this book, we thank for their valuable contributions and constructive feedbacks. It is a great honour to work with this great team of wonderful friends and experts in the area of this book.

We wish to thank the publisher, Springer-Verlag, for their excellent support during the preparation of this manuscript. Finally, we are very grateful to the Conicyt Fondef Project D10I-1198, Innova Corfo Project 11DL2-10399 and the Chilean Millennium Institute of Complex Engineering Systems (ICM: P-05-004-F, CONICYT: FBO16), which partially supported this endeavour.

Santiago, Chile                                              Juan D. Velásquez
Oxford, United Kindom                                          Vasile Palade
Adelaide, Australia                                          Lakhmi C. Jain
July 2012

# Contents

# List of Contributors

Pablo E. Román
Web Intelligence Consortium Chile Research Centre, Department of Industrial
Engineering School of Engineering and Science, University of Chile,
Av. República 701, Santiago, Chile, P.C. 837-0720
e-mail: proman@ing.uchile.cl

Gastón L'Huillier
Groupon, Inc.,
3101 Park Blvd., Palo Alto, CA. 94301, USA
e-mail: gaston@groupon.com

Pablo Loyola Heufemann
Division of Web Science and Technology, Korea Advanced Institute of Science &
Technology,
Web Engineering and Adaptive Software Laboratory 335 Gwahangno (373-1
Guseong-dong), Yuseong-gu, Daejeon 305-701, KI Building, 3rd floor, room C304,
Republic of Korea
e-mail: ployola@kaist.ac.kr

Edison Marrese Taylor
Web Intelligence Consortium Chile Research Centre, Department of Industrial
Engineering School of Engineering and Science, University of Chile,
Av. República 701, Santiago, Chile, P.C. 837-0720
e-mail: emarrese@wi.dii.uchile.cl

Cristián Rodríguez O.
Web Intelligence Consortium Chile Research Centre, Department of Industrial
Engineering School of Engineering and Science, University of Chile,
Av. República 701, Santiago, Chile, P.C. 837-0720
e-mail: crodriguezo@wi.dii.uchile.cl

Goldina Ghosh
Department of Computer Science,
Birla Institute of Technology,
Mesra, India
e-mail: goldinag@gmail.com

Soumya Banerjee
Department of Computer Science,
Birla Institute of Technology,
Mesra, India
e-mail: dr.soumya@ieee.org

Giovanna Castellano
University of Bari "A. Moro", Department of Informatics,
Via Orabona, 4 - 70125 Bari (Italy)
e-mail: castellano@di.uniba.it

Anna M. Fanelli,
University of Bari "A. Moro", Department of Informatics,
Via Orabona, 4 - 70125 Bari (Italy)
e-mail: fanelli@di.uniba.it

Maria A. Torsello,
University of Bari "A. Moro", Department of Informatics,
Via Orabona, 4 - 70125 Bari (Italy)
e-mail: torsello@di.uniba.it

Jorge Gaete Villegas
Department of Computer Science, Korea Advanced Institute of Science &
Technology,
291, Daehak-ro, Yuseong-gu,Daejeon 305-701, Republic of Korea
e-mail: jorge@kaist.ac.kr

In-Young Ko
Department of Computer Science and Division of Web Science and Technology,
Korea Advanced Institute of Science & Technology 291, Daehak-ro, Yuseong-
gu,Daejeon 305-701, Republic of Korea
e-mail: iko@kaist.ac.kr

Denis Parra
School of Information Sciences, University of Pittsburgh
135 North Bellefield Avenue, Pittsburgh, PA 15260, USA
e-mail: dap89@pitt.edu

Shaghayegh Sahebi
Intelligent Systems Program, University of Pittsburgh
Sennott Square, Pittsburgh, PA 15260, USA
e-mail: ssahebi@cs.pitt.edu

Vasile Palade,
Department of Computer Science, University of Oxford, UK,
Wolfson Building, Parks Road, Oxford OX1 3QD, UK
e-mail: vasile.palade@cs.ox.ac.uk

Lakhmi C. Jain,
KES Centre, School of Electrical and Information Engineering,
University of South Australia, Adelaide, Mawson Lakes Campus,
South Australia SA 5095, Australia
e-mail: Lakhmi.Jain@unisa.edu.au

Juan D. Velásquez
Web Intelligence Consortium Chile Research Centre, Department of Industrial
Engineering School of Engineering and Science, University of Chile,
Av. República 701, Santiago, Chile, P.C. 837-0720
e-mail: jvelasqu@dii.uchile.cl

# Chapter 1
# New Trends in Web User Behaviour Analysis

Pablo E. Román, Juan D. Velásquez, Vasile Palade, and Lakhmi C. Jain

**Abstract.** The analysis of human behaviour has been conducted within diverse disciplines, such as psychology, sociology, economics, linguistics, marketing and computer science. Hence, a broad theoretical framework is available, with a high potential for application into other areas, in particular to the analysis of web user browsing behaviour. The above mentioned disciplines use surveys and experimental sampling for testing and calibrating their theoretical models. With respect to web user browsing behaviour, the major source of data is the web logs, which store every visitor's action on a web site. Such files could contain millions of registers, depending on the web site traffic, and represents a major data source about human behaviour. This chapter surveys the new trends in analysing web user behaviour and revises some novel approaches, such as those based on the neurophysiological theory of decision making, for describing what web users are looking for in a web site.

## 1.1 Introduction

Since early times of civilization, humanity has faced the challenge of understanding itself. Traders anticipate peoples' needs, politicians calculate the move with

Pablo E. Román · Juan D. Velásquez
Web Intelligence Consortium, Chile Research Centre,
Department of Industrial Engineering School of Engineering and Science,
University of Chile, Av. República 701, Santiago, Chile, P.C. 837-0720
e-mail: proman@ing.uchile.cl, jvelasqu@dii.uchile.cl

Vasile Palade
Department of Computer Science, University of Oxford, UK,
Wolfson Building, Parks Road, Oxford OX1 3QD, UK
e-mail: vasile.palade@cs.ox.ac.uk

Lakhmi C. Jain
KES Centre, School of Electrical and Information Engineering, University of
South Australia, Adelaide, Mawson Lakes Campus, South Australia SA 5095, Australia
e-mail: Lakhmi.jain@unisa.edu.au

J.D. Velásquez et al. (Eds.): Advanced Techniques in Web Intelligence-2, SCI 452, pp. 1–10.
springerlink.com                    © Springer-Verlag Berlin Heidelberg 2013

the best political outcome and generals decide the position of the army. Human beings live together in societies constituting complex systems of interdependence. One stepping stone for the construction of a better society consists in having sufficient knowledge of human behaviour [25].

Social science has recently used many modern tools like dynamic and stochastic systems for describing the structures of social change [4]. Every social system has naturally been described as a highly complex network of interaction that has been considered impossible to represent mathematically. Nevertheless, with the help of abstraction many models have been designed for explaining particular social phenomena. For example, applications to politics are revealed by the modeling of the mass's opinion dynamics. A model for voter dynamics [16] could help to understand the mass's voting intentions.

In a business environment, marketing is the set of processes that helps to determine customer preferences and demand for products and services, recommending strategies for sales and communications. Google is a company that bases its main business on selected publicity on a search result page. Google's marketing strategy is based on web users' preference rankings for web pages. A simple stochastic model for web user browsing (Page Rank algorithm [14]) generates stationary probabilities for calculating this ranking. Therefore, the Internet has become a new frontier in the marketing field that promises commercial success, but at the cost of the need to have accurate knowledge about the web user [25].

Some successful examples in e-commerce, such as Amazon.com, could be mentioned. Amazon is a USA company that is mainly designed as an online book store, but has also moved into trade in electronic devices, furniture and other goods. Amazon.com is considered to be the first promoter of online shopping, with prosperous business results. Its annual net income was about US\$ 902 million in 2009. Amazon relies on online recommendations to customers according to their detected pattern profiling. Such technologies are based on predicting the preferences of a web user based on his/her observed navigational behaviour.

Netflix is another dot-com company with US\$ 719 million net income in 2011, which focuses on DVD movie rental. In 2006, this company offered a one million-dollar contest for improving by ten percent the effectiveness of its movie-recommending algorithm. However, the prize was only claimed in September 2009, after roughly four years of worldwide competitors' attempts [10]. The winner used a specific data mining algorithm, [27] similar to many others. The main conclusion that can be drawn is that modeling human behaviour is a very difficult problem.

## 1.2   The Web Operation

In order to study web user navigational behaviour, it will be important to clarify some basic notions about the web system first (Figure 1.1). Web users are considered human entities that, by means of a web browser, access information resources in a hypermedia space called the World Wide Web (WWW). Common web users'

objectives are information foraging (looking for information about something), social networking activities (e.g., Facebook), e-commerce transactions (e.g., Amazon
Shopping), bank operations, etc. On the other hand, the hypermedia space is organized into web pages that can be described as perceived compact subunits called
"web objects". The design of web pages is created by "web masters" who are in
charge of a group of pages called a "web site." Therefore, the WWW consists of a
vast repository of interconnected web sites for different purposes.

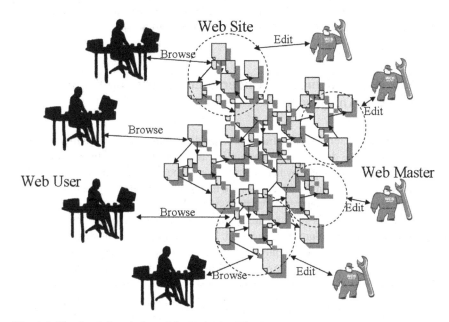

**Fig. 1.1** First level description of the Web User/Site System

On a second level of abstraction, web pages are distributed by "web servers,"
and "web browsers" are used to trigger the mechanism. A web page is an HTML-
encoded document, which contains hyperlinks to other pages. The content of a web
page corresponds to the visual text and multimedia that a web user perceives when
the browser interprets the HTML code. The web structure corresponds to a graph
of pages interconnected by hyperlinks. Actually, both the content and the structure have been highly dynamic since the web 2.0 application began personalizing
web sites to the current web user, like Facebook or Amazon. In this context, the
system, comprised by web users that navigate on web sites, has a highly complex
description.

The framework in [3] was the first application of a model of human behaviour
on the web, like the Page Rank algorithm [14]. This algorithm has had a very big
reputation, since it was used by the Google search engine. It anticipates the user
visiting any link on the page with equal probability. This stochastic model also includes the probability of restarting the process by recommencing the navigation in

another uniformly distributed page, which results in an ergodic Markov chain. This simple stochastic process adheres to stationary probability rules. The most important or interesting pages have the highest probability of a random user visiting it. This process creates a ranking for the pages used in web search engines.

Further refinements of this idea were used in [21], where web users are considered as flows over the network of hyperlinks. Other approaches relate to models for evaluating the navigability of web sites. In particular [28], incorporates common browsing actions that a web user performs, such as: terminating a session, proceeding to, going back, staying and jumping to. Each one of these actions relates to a probability value that is incorporated into a Markov chain model.

## 1.3   The Big Challenge: To Understand the Human Behaviour by Data Analysis

Data mining is the automated extraction of predictive information from generally large databases. This definition has an implicit assumption that a statistical methodology is used. Data mining is by nature proactive, in the sense that the user must tune the algorithm, because, in the process, some unexpected pattern may be discovered, which the miner has to interpret. Nevertheless, much of the prospective effort can be automated. Data mining is an automated pattern discovery on data, and for such purposes machine learning algorithms are used. Data mining, in fact, uses generic tools and does not rely on the internal physical processes from which data are generated. However, once patterns are discovered, multiple selection algorithms can be used for prediction. Successful applications have been performed, such as credit scoring, weather prediction, revenue management, customer segmentation, and many others.

Nevertheless, data mining is not a silver bullet for many problems including human behaviour analysis. The hierarchy cascade data mining process [8] has shown that many human-controlled iterations are needed for finally adjusting a model to data and to applying it in a predictive mode. Nowadays, new computer improvements allow having a more automatic process for adjusting machine learning models, which make up the new generation of intelligent applications. An editorial in the Wired magazine, entitled "The End of Theory: The Data Deluge Makes the Scientific Method Obsolete" claims that current intelligent applications are quite powerful enough to handle any given real-world complexity, making many theories obsolete [2]. It is obvious that the explosive increase of data had an impact on the precision of many modeling techniques. Hence, this proposal seems too optimistic to be true, and there are fundamental reasons for discarding this extremely naive proposal.

First, automatic machine learning methods improve the accuracy of predictions by training better the models on the available data. The problem is that future scenarios need to belong to the same kind of data. For instance, if we try to model the trajectory of the Apollo mission by a machine learning approach, perhaps the model will perform well. But if a new factor like cosmic dust enters into the scenario, such a machine learning model will fail. On the other hand Newton's theory does not change in this case and the new factor can be included as a viscosity term on the theoretical model. Second, the large quantities of data presented here do not compare with the large quantities of data that could be extracted from other examples in nature. A simple fluid motion in a box contains an Avogadro number of particles ($N = 10^{23}$) and a much larger number of possible configurations ($N!$), therefore the problem is that, for a computer to be capable of handling such an amount of information, it would have to be constructed of a number of memory pieces much larger than the estimated number of atoms in the universe. Of course, fluid theory can model the average motion of particles in a fluid, and this is the best-known way to handle this problem. Third, intractable computational complexity is easy to find in a simple human activity system, like the traveling salesman problem, which belongs to the class of NP-Hard problems. Nevertheless, there are several heuristics based on machine learning methods, but they show average performance in some cases and worse performance in others.

The Internet has become a regular channel for communication, most of all for business transactions. Commerce over the Internet has grown to higher levels in recent years. For instance, e-shopping sales have been dramatically increasing recently, achieving a growth of 17% in 2007, and generating revenues of $240 billion/Year in the US alone [5]. This highlights the extent of the importance of acquiring knowledge on how the Internet monitors customer's interactions within a particular web site [24, 25].

One can analyze this new technological environment using traditional marketing approaches, but the Internet invites new methods of determining consumer's genuine needs and tastes. Traditional market surveys serve no purpose in reflecting the actual requirements of customers who have not been precisely defined in the web context. It is well known that web users are ubiquitous. In that sense, a marketing survey compiled in a specific location in the world does not carry clear statistical significance in another. However, online queries should improve this issue by requesting that each visitor answers as many focused questions as they can [7], but apart from predicting future customer preferences, online surveys can improve the effectiveness of the web site content strategy.

According to the WIC (Web Intelligence Consortium), the term "Web Intelligence" corresponds to "Research and development to explore the fundamental roles and impact of Artificial Intelligence and advanced Information Technology on the next generation of web-empowered systems, services, and activities." In this context, Web usage mining (WUM) can be defined as the application of machine learning techniques on web data for the automatic extraction of behavioural patterns of web users. In this sense, web usage patterns can be used for analyzing web user preferences. Traditional data mining methods need to be pre-processed and adapted before being employed on web data. Several efforts have been made to improve the quality of the results. Once a repository of web user behaviour (Web Warehouse) is available [26], specific machine learning algorithms can be applied in order to extract patterns regarding the usage of the web site. As a result of this process, several applications can be implemented on adaptive web sites, such as recommender systems and revenue management marketing, among others.

## 1.4   Novel Approaches for Modeling the Web User Behaviour

While current approaches for studying the web user's browsing behaviour are based on generic machine learning and data mining approaches [17], a rather different point of view has been developing in the past decade. New models based on the neurophysiology theory of decision making have been applied to the link selection process. These models have two stages: training and simulation. In the first stage, the model's parameters are adjusted to the user's data. In the second one, the configured agents are simulated within a web structure for recovering the expected behaviour. The main difference with the machine learning approach consists in the model being independent of the structure and content of the web site. Furthermore, agents can be confronted with any page and decide which link to follow (or leave the web site). This important characteristic makes these kind of models appropriate for heavily dynamic web sites. Another important difference is that these models have a strong theoretical basis built upon a physical phenomenon, whose modeling equations came from a phenomenon observation. Traditional physics models approaches are more generic, but, for web user behaviour analysis, the proposal is based on a specific state-of-the-art theory of brain decision making.

New web user behavior models are related to the brain neural activity levels (NAL) of certain brain regions with a discrete set of possible choices. For instance, the LCA (Leaky Competing Accumulator) [23] is used for analyzing the NALs ($X_i$) evolution according to a stochastic equation during the agent's decision making process until one of the NAL's values reaches a given threshold. It describes the neural activity of different brain regions during the subject resolution of the decision, by means of a stochastic process that evolves until an activity reaches a given threshold that fires the decision. Such a class of stochastic processes applied to decision making has been experimentally studied for nearly forty years [9, 15, 19, 12]. In this context, a web user is confronted with deciding which link to follow according to his/her own purposes, and the process is repeated again for each visited page until leaving the web site. Then the web user faces a set of discrete decisions that corresponds to the selection of a hyperlink (or leaving the site) as in figure 1.2.

Here, the LCA model simulates the artificial web user's session by estimating the user's page sequences and, furthermore, by determining the time taken in selecting an action, such as leaving the site or proceeding to another web page. Experiments performed using artificial agents that behave in this way highlight the similarities between artificial agents' results and a real web user model of behaviour. Moreover, the performance of the artificial agents is reported to have similar statistical behaviour to humans. If the web site semantic does not change, the set of visitors remains the same. This principle enables predicting the changes in the access pattern to web pages related to small changes in the web site that preserve the semantic. The web user's behaviour could be predicted by simulation and then services could be optimized. Other studies on ant colony models [1] relate directly to general purpose clustering techniques.

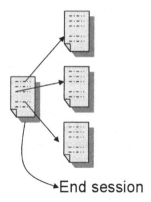

**End session**

**Fig. 1.2** Navigational options on a web page

Such an analysis of web user behaviour requires a non-trivial data pre-processing stage, in order to obtain the sequence of web pages (session) for individual visitors, the text content and the hyperlink structure of the web site. Novel algorithms are developed, based on integer programming, and are used for the optimal extraction of web users' sessions [13] . Traditionally, the next step consists of applying data mining techniques for identifying and extracting patterns of web user browsing behaviour. Furthermore, data quality needs to be ensured, since the calibration of web user models is sensitive to the data set.

## 1.5   Characterizing the Web User Browsing Behaviour

As described in [6, 11, 18, 20], web usage data can be extracted from different sources, with web logs being the main resource for web usage mining applications. The variety of different sources carries a number of complexities in terms of data pre-processing, for example related to the incompleteness of each source. As a solution to this, several pre-processing algorithms have been developed [26]. Further sources, like the hyperlink structure and the web page content, complement the extracted log information, providing a semantic dimension of each user's action. Furthermore, in terms of the web log entries, several problems must be confronted. Overall, a web log in itself does not necessarily reflect a sequence of an individual user's documented access. Instead, it registers every retrieval action, but without a unique identification for each user.

The web user browsing behaviour can be described by three kinds of data: the web site structure, the web page content and the web user session. The first is directly related to the environment. The third describes the click stream that each Web user performs during his visit to the web site.

## 1.6   Real-World Applications

Current Web 2.0 applications [22] are highly popular on the Web and have enormous potential for commercial use. Nevertheless, the analysis of web usage is ob-
fuscated by the dynamic characteristic of the content and structure of the web sites.
Today's applications of web mining have been fully explored in terms of both fixed
content and structure, with the help of natural language processing techniques and
specialized machine learning algorithms. A web recommender system is a typical
web intelligence application [24]. Recommendation is based on a web user profile
recognition procedure, which filters information for content on a web page. Such a
system is based on detected common patterns in the web user browsing behaviour.

## 1.7   Chapters Included in the Book

This book contains seven chapters contributed by several known researchers and
experts in the Web User Behaviour Analysis area. In a broad perspective, this book
includes topics such as traditional and new web data pre-processing techniques,
approaches for modelling the web user inspired from the neuro-science area, web
usage mining algorithms and techniques, extraction of relevant user opinions from
web 2.0 applications, web adaptive and recommendation systems.

Chapter two, "Web Usage Data Pre-processing", by Gastón L'Huillier and Juan
D. Velásquez, presents different web data collection strategies, as well as merging
and cleaning techniques for web usage data, considering the data originated in the
Web and also new data generated trough the application of new tools, like Eye tracking systems.

Chapter three, "Cognitive science for web usage analysis" by Pablo E. Román,
and Juan D. Velásquez, addresses different techniques in cognitive science and
adapts them for the understanding of human behaviour, in the particular case of
the web user behaviour analysis.

Chapter four, "Web Usage Mining: Discovering Usage Patterns in Web Applications" by Giovanna Castellano, Anna M. Fanelli and Maria A. Torsello, introduces a
novel WUM approach based on the use of fuzzy clustering to the discovery of user
categories, starting from usage patterns.

Chapter five, "Web Opinion Mining and Sentimental Analysis" by Edison Marrese Taylor, Cristián Rodríguez O., Juan D. Velásquez, Goldina Ghosh and Soumya
Banerjee, introduces theories and techniques for extracting knowledge from the user
opinions in web 2.0 systems, and shows a real-world application in Twitter.

Chapter six, "Web Usage Based Adaptive Systems", by Pablo Loyola Heufemann, Jorge Gaete Villegas and In-Young Ko, presents different techniques and issues in the area of user-adaptive software systems applied to Web environments
and proposes a set of directions for the future development of Web Usage Based
Adaptive Systems in the new Internet environments.

Chapter seven, "Recommender Systems: Sources of Knowledge and Evaluation Metrics", by Denis Parra and Shaghayegh Sahebi, presents a classification of web recommender systems and the main sources of knowledge and evaluation metrics that have been described in the research literature in the area.

# References

1. Abraham, A., Ramos, V.: Web usage mining using artificial ant colony clustering and genetic programming. In: Procs. of the 2003 IEEE Congress on Evolutionary Computation (CEC 2003), pp. 1384–1391 (2003)
2. Anderson, C.: Wired Magazine, Editorial (June 2008)
3. Blum, A., Chan, T.-H.H., Rwebangira, M.R.: A random-surfer web-graph model. In: Proceedings of the Eigth Workshop on Algorithm Engineering and Experiments and the Third Workshop on Analytic Algorithmics and Combinatorics. Society for Industrial and Applied Mathematics, pp. 238–246 (2006)
4. Castellano, C., Fortunato, S., Loreto, V.: Statistical physics of social dynamics. Reviews of Modern Physics 81(2), 591 (2009)
5. Grannis, K., Davis, E.: Online sales to climb despite struggling economy. According to Shop.Org/Forrester Research Study (2008)
6. Jin, X., Zhou, Y., Mobasher, B.: Web usage mining based on probabilistic latent semantic analysis. In: KDD 2004: Proceedings of the Tenth ACM SIGKDD International Conference on Knowledge Discovery and Data Mining, pp. 197–205. ACM, New York (2004)
7. Kausshik, A.: Web Analytics 2.0: The Art of Online Accountability and Science of Customer Centricity. Sybex (2009)
8. Kosala, R., Blockeel, H.: Web mining research: A survey. SIGKDD Explorations: Newsletters of the Special Interest Group (SIG) on Knowledge Discovery and Data Mining 1(2), 1–15 (2000)
9. Laming, D.R.J.: Information theory of choice reaction time. Wiley (1968)
10. Lohr, S.: A 1 million dollars research bargain for netflix, and maybe a model for others. New York Times (2009)
11. Masseglia, F., Poncelet, P., Teisseire, M., Marascu, A.: Web usage mining: extracting unexpected periods from web logs. Data Min. Knowl. Discov. 16(1), 39–65 (2008)
12. Ratcliff, R.: A theory of memory retrieval. Psychological Review (83), 59–108 (1978)
13. Romn, P.E., Dell, R.F., Velsquez, J.D., Loyola, P.: Identifying user sessions from web server logs with integer programming. Intelligent Data Analysis (to appear, 2012)
14. Brin, S., Page, L.: The anatomy of a large-scale hypertextual web search engine. In: Computer Networks and ISDN Systems, pp. 107–117 (1998)
15. Schall, J.D.: Neural basis of deciding, choosing and acting. National Review of Neuroscience 2(1), 33–42 (2001)
16. Schneider-Mizell, C.M., Sander, L.M.: A generalized voter model on complex networks. Technical Report arXiv:0804.1269, Department of Physics, University of Michigan, 15 pages, 3 figures (April 2008)
17. Sebastiani, F.: Machine learning in automated text categorization. ACM Comput. Surv. 34(1), 1–47 (2002)
18. Srivastava, J., Cooley, R., Deshpande, M., Tan, P.: Web usage mining: Discovery and applications of usage patterns from web data. SIGKDD Explorations 2(1), 12–23 (2000)
19. Stone, M.: Models for choice reaction time. Psychometrika (25), 251–260 (1960)

20. Tao, Y.-H., Hong, T.-P., Lin, W.-Y., Chiu, W.-Y.: A practical extension of web usage mining with intentional browsing data toward usage. Expert Syst. Appl. 36(2), 3937–3945 (2009)
21. Tomlin, J.A.: A new paradigm for ranking pages on the world wide web. In: WWW 2003, Budapest, Hungary, May 20-24 (2003); In: Computer Networks and ISDN Systems, pp. 107–117 (1998)
22. Ullrich, C., Borau, K., Luo, H., Tan, X., Shen, L., Shen, R.: Why web 2.0 is good for learning and for research: principles and prototypes. In: WWW 2008: Proceeding of the 17th International Conference on World Wide Web, pp. 705–714. ACM, New York (2008)
23. Usher, M., McClelland, J.: The time course of perceptual choice: The leaky, competing accumulator model. Psychological Review 2(1), 550–592 (2001)
24. Velasquez, J.D., Palade, V.: Building a knowledge base for implementing a web?based computerized recommendation system. International Journal of Artificial Intelligence Tools 16(5), 793–828 (2007)
25. Velasquez, J.D., Palade, V.: A knowledge base for the maintenance of knowledge extracted from web data. Knowledge?Based Systems Journal 20(3), 238–248 (2007)
26. Velásquez, J.D., Palade, V.: Adaptive web sites: A knowledge extraction from web data approach. IOS Press, Amsterdam (2008)
27. Wolpert, D.H.: Stacked generalization. Neural Networks 5, 241–259 (1992)
28. Zhou, Y., Leung, H., Winoto, P.: Mnav: A markov model-based web site navigability measure. IEEE Trans. Softw. Eng. 33(12), 869–890 (2007)

# Chapter 2
# Web Usage Data Pre-processing

Gaston L'Huillier and Juan D. Velásquez

**Abstract.** End users leave traces of behavior all over the Web all times. From the explicit or implicit feedback of a multimedia document or a comment in an online social network, to a simple click in a relevant link in a search engine result, the information that we as users pour into the Web defines its actual representation, which is independent for each user. Our usage can be represented by different sources of data, for which different collection strategies must be considered, as well as the merging and cleaning techniques for Web usage data. Once the data is properly pre-processed, the identification of an individual user within the Web can be a complex task. Understanding the whole life of a user within a session in a Web site and the path that was pursued involves advanced data modeling and a set of assumptions which are modified every day, as new ways to interact with the online content are created. The objective is to understand the behaviour and preferences of a web user, also when several privacy issues are involved, which, as of today, are not clear how to be properly addressed. In this chapter, all previous topics regarding the processing of Web usage data are extensively discussed.

## 2.1 Introduction

Let us think about our fingerprints and how they are everywhere in our surroundings. The things we touch everyday are a living record of the tools we use for getting something done. Imagine we could keep a record of each one of the objects we

Gaston L'Huillier
Groupon, Inc., 3101 Park Blvd., Palo Alto, CA. 94301, USA
e-mail: gaston@groupon.com

Juan D. Velasquez
Department of Industrial Engineering, University of Chile,
Republica 701, Santiago, Chile
e-mail: jvelasqu@dii.uchile.cl

J.D. Velásquez et al. (Eds.): Advanced Techniques in Web Intelligence-2, SCI 452, pp. 11–34.
springerlink.com

touch in a regular day and its potential outcome. This information could lead us to understand which are the most relevant tools and objects in our everyday life and would reveal several highly personal properties of ourselves. Within the Web, with the current technology available, the analogy is something that you can consider as real. The links we click, the videos we play, the images we look at, and even the text we write, are living records of our most personal interests. These fingerprints we pour everyday into the Web can be translated into models which can be used to understand a brand's customers, the best places to have dinner tonight in the city you live according to other customers, the most relevant results when searching for something, the opinions on political events, and even the general reaction to a viral video.

Once you understand the user, you can provide a better interaction and experience with your business. The holy grail of personalization is having a complete understanding of which are the most relevant topics and elements for a particular user, in order to maximize the added value when delivering a service. This way, whether it is serving the user with the best online catalog with items he might like, or the best possible links when searching in a search engine, the user will come back, and will use your service. However, extreme personalization might not be considered a good thing, as some may argue that people need to know the elements which are beyond their surrounding world. To provide a user with choices and still allow the possibility of diversity is something equally important to make the business grow, and to push ourselves to know beyond our own boundaries and comfort zones.

To be able to extract the end user's behavior and interests, it is necessary to understand the nature of the data available, how to collect it, and how to process it. In this chapter, we examine various concepts related to the sources and collection techniques for Web usage data, and a general overview of the most relevant methods for Web usage data processing is presented. Methods and techniques presented in this chapter are what we consider as relevant for the particular examples discussed, and are not intended to include an exhaustive survey of equally important methods which might be considered for other applications.

In this chapter, Web usage data sources and collection techniques are presented in Section 2.2. Then, in section 2.3 methods for Web session reconstruction and user identification are introduced. Profiling methods and methods for processing relevance data such as explicit or implicit feedback from the Web users are reviewed in Section 2.4. Finally, privacy issues are discussed in Section 2.5.

## 2.2 Web Usage Data Sources and Collection

As in any other data mining task for Web usage mining, the collection of useful data is fundamental. This is reflected by the fact that many pattern recognition methods are supported by the "garbage-in, garbage-out" principle. Moreover, the information for Web usage mining is often ruled by large data collection systems closely related to a system or platform used by several end users. As these users might not be aware of the real purposes of the system, their behavior can be erratic and unintentionally go against for what it was designed for. To identify and understand which elements are not part of the actual interests of the user is a responsibility for both the practitioner and the researcher. Practitioners have to be conscious of what the objectives of a given website are, and build it in terms of clear use cases. Likewise, researchers need to understand which were the original objectives from which the data was generated, and model the interaction with the user accordingly. If any of these duties are overlooked and not taken into consideration, the collection and processing of Web usage data could lead to uncertain and misleading results.

Web data has mainly three representations, the *Web Structure Data*, *Web Content Data*, and the *Web Usage Data* [49]. All these types of Web data are useful when trying to determine the actual usage of the Web site or application.

Web structure data corresponds to the underlying hyperlink structure that Web documents have. This structure has been historically considered as static, but as the interaction between Web documents has gained a dynamic interaction, the Web structure data has evolved. This structure when defined by explicit hyperlinks within a given Web page, can be directly extracted and used for further analysis. However, this is generally not the case as different frameworks within programming languages dynamically pre-build the links between different Web documents, and which can be modified by the end users as they interact with the website.

Web content data is associated with all the information that the end user consumes when accessing a particular Web page. In this case, the content data is stored in Web documents which are generally translated into HTML documents, multimedia documents (e.g. images, videos, and audio files), and files with proprietary or open formats (e.g. Portable Document Format (PDF), MS Word (DOC), MS Excel (XLS), among other formats used). Furthermore, Web content documents can be separated in two main categories, documents that contain the information that the end user visualizes when navigating a particular Web page or application, and documents that are associated with the metadata that the Web document could have. Take for example, semantic information translated into micro-formats [42] or any kind of data that describes the actual information that the user is accessing. In some cases, the documents accessed by the Web user have both types of data in a single document, and sometimes there are independent documents related by hyperlinks to store further semantic definitions.

Web usage data is that related to the activity of Web users within a particular website. This data is generally composed of a sequence of pageviews stored in Web access logs which traces all the requests and click-through data related to a given user. In the following, a further description of this type of data is presented with a general discussion on its processing problems.

## 2.2.1  Nature of Web Usage Data

The sequence of pageviews associated with a particular user when navigating a Website can be extracted and stored by several methods, and the decision on which methods should be used depends whether the usage data is related to user authentication methods or not. One method that has been widely used and is popular because of its severe privacy implications is the usage of explicit tracking devices that a Web server installs in the user's browser (commonly known as *Web browser cookies*). These devices are small lines of code that store all the user information from the web browser, and are accessed by the server that installed the device. It has been reported that this information can be shared with third parties for advertisement analysis and other applications [17]. Also, installing these tracking devices is equivalent to installing spyware. However, less intrusive methodologies to rebuild a particular user session have been proposed. These methods will be reviewed in depth in Section 2.3.

As of today, the most common Web server used by practitioners and developers is the Apache HTTP server[1]. Also, open source Web application servers like NGinX[2] or proprietary servers like Microsoft IIS[3] are commonly used by developers to serve Web pages over the Web[4]. Even though there is a standard defined by the World Wide Web Consortium, some of servers might have different properties and options for storing the Web access logs. The collection of these logs is related to the server configuration in production servers, determined by the system administrators of a particular Website or application. However, standard formats such as the Common Log Format CLF[5] defines the basic elements that a particular record will have in a log.

To perform a large scale analysis over the Web access logs, several steps have to first be considered. For example, due to the volume that these logs can achieve, access to them as well as the processing capabilities are usually constrained and must be considered as a process to be executed in independent servers separated from the original server where the access logs are originally stored.

Once the data is collected from the different sources, the deletion of unnecessary requests is fundamental to reduce the volume of data to be analyzed. This has to be realized by first determining which HTTP requests do not contribute valuable information for the study that is being performed. HTTP requests are determined by a status code which is standard and universally adopted by regular HTTP Web servers[6].

---

[1] http://httpd.apache.org/ [last accessed April 18, 2012].

[2] http://nginx.org/ [last accessed April 18, 2012].

[3] http://www.iis.net/ [last accessed April 18, 2012].

[4] In terms of active sites, Apache has a 58.21%, MS ISS a 12.31%, and NGinx a 11.61%, for more information refer to http://bit.ly/WebServersActiveSites [last accessed April 18, 2012].

[5] For more information on the Web log format, please refer to http://www.w3.org/Daemon/User/Config/Logging.html [last accessed April 18, 2012].

[6] For further details on the requests status codes, check the W3C description included in the RFC 2616: http://www.w3.org/Protocols/rfc2616/rfc2616-sec10.html [last accessed April 18, 2012].

An entry log consists of the user's IP address, the access time, the request method, the URL of the page accessed, the protocol, the return code, the user-agent, and the number of bytes transmitted. Basic filtering rules are used by practitioners and researchers when processing this information. Among these rules, the removal of any results in error, request methods different than GET, and records from multimedia (e.g. images) file access are the most commonly used. While error logs might be interesting for some types of analysis and debugging strategies for developers, it is unusual to consider this type of information when extracting useful patterns related to the preferences and behavior of end users.

### 2.2.2 Merging and Cleaning Data

Usually, the data related to Web log activities is distributed in several servers, which includes multiple instances of the same data, used as a strategy to reduce server loads when being accessed by the end users. All this redundant information has to be merged accordingly, and this process requires the synchronization of logs for which the usage of heuristics considers data from sessionization and user identification methods [27].

As mentioned in Section 2.2.1, each time a Web page is accessed, the Web server registers all the information related to the website, like style pages and references to embedded objects, such as multimedia files. This non-relevant information has to be removed from the Web logs as it does not contribute useful information to any further analysis. Also, depending on the application and the analysis that is being performed, some fields from the log might not be useful (e.g. bytes transferred, HTTP protocol version, etc.) and its removal might cause a decrease in the data volume to be analyzed.

Besides removing all information which is not relevant for the given application and interests of the researcher and/or practitioner, it is highly important to remove all the data that was not generated by actual Web users. Such data could be associated with Web crawler activities that register non-relevant information into the Web server logs and/or application server logs. For the removal of Web crawler entries in the access logs, a list of well-known crawlers can be used to filter the non-relevant records. Some well-behaved Web crawlers log their user-agent before generating logs in a particular website (usually in a file named "robots.txt"), which is a useful resource for data-cleaning purposes. However, there are cases when the crawler is from an unknown source and it does not identify itself as a Web crawler. For dealing with this type of Web crawlers, different classification strategies have been proposed [8, 41, 43].

## 2.3 Web Session Reconstruction and User Identification

Sessionization can be defined as the process of identifying a particular user session from Web data. This process is a basic component for the success of any Web usage mining application. Sessions could be determined explicitly by a user when it is logged in the application, or determined by heuristics methods when there is no explicit knowledge on who is the Web user, and how a particular pageview is related to them [46, 47].

In this section, questions on how to build sessions out of Web data when there is no explicit correlation with a particular Web user are reviewed. A brief introduction on strategies to deal with incomplete information cases, where there is insufficient data to build a session is presented. Finally, a review on performance metrics used for the model assessment on sessionization methods is discussed.

## 2.3.1   Heuristics and Methods for Sessionization

To build the sessions from Web server logs, the most common approaches are the *proactive* and *reactive* methods. On the one hand, proactive methods retrieve directly from the Web user the information to determine the pageviews related to the particular session. This information can be retrieved by a cookie enabled on the client/browser of the user, or by the authentication of the end user, where each click is then directly associated with the actual user. On the other hand, the data might not be that easy to retrieve, given privacy issues or the refusal of the Web user to accept the conditions and usage terms of the application, and there is no authentication. In this case, in order to build the user's session and to determine which pageviews are related to each other, the analysis must be performed using the Web server logs.

When dealing with Web server logs for the processing, once the data is cleaned and merged, heuristics to determine the sessions for each one of the users have to be performed. These heuristics range from basic methods that take IP addresses, user-agents and time windows to create a session, to advanced algorithms which consider all the previous information as input to machine learning or optimization methods which try to adjust the best sequence of pageviews for a single user session. Basic heuristics are easy to develop and run faster than advanced methods, but have a very low accuracy as many users might be coming from the same IP address and user-agent combination. Advanced methods yield a higher accuracy, but might have a high computational complexity in both time execution and disk storage.

### 2.3.1.1   Session Representation

Different representations can be used for a user session [7], and which representation will be used is closely related to the techniques and actual application that is going to be performed. The most common representations are weighted usage per page and text weight representation.

According to Mobasher et al. in [29], sessions can be processed to obtain a weight vector $s = [w_1, \ldots, w_n]$ where each component is a Web page normalized according to the time of visit. In this case, $n$ represents the total number of Web pages on a particular website. Vector entries could be defined as null or zero when the Web page is not visited during a session, and related to its time of visit whenever the Web page was visited by Web users. In the simplest case, as proposed in [28], the representation can be defined as binary, where $w_i \in \{0, 1\}$. Both binary and continuous definitions can be used for different input algorithms. For example, the binary representation has been reported to be useful for algorithms based on association rules [28] and the continuous representation for clustering methods [29].

Another approach introduced in [12], is related to the order in which the sessions were accessed $s =< t_i, (p_{t_i}^1, \ldots, p_{t_i}^n), (T_{p_{t_i}^1}, \ldots, T_{p_{t_i}^n}) >$, where $t_i$ is a unique session identifier, $p_{t_i}^1, \ldots, p_{t_i}^n \in P$ is the ordered user requests for session $t_i$, $T_{p_{t_i}^1}, \ldots, T_{p_{t_i}^n}$ is the normalized time values for requests in $P$, and $P$ is a set with all available requests. As obvious as it seems, to consider as a component the actual order of the Web pages lead to further analytical properties which are not covered in previous weighted representation. For example, similarity measures can be computed considering the sequence as an adjacency matrix based on pair-wise scores that lead to a graph representation of two sessions. This representation can be used to define several distance scores, which leads to the construction of several clustering methods [12].

Furthermore, as presented in [49], an approach which considers an ordered text weight representation is taken into consideration. Here, text properties and semantics are blended with the order of the pageviews and the relative time spent by the user in each Web page. Furthermore, this can be generalized for multimedia files using Web objects definitions as presented in [48]. Clustering methods such as Self-Organized Feature Maps (SOFM) [23] can be employed using this representation to reveal special Web usage properties from the data that is being processed, such as determining which are the website key objects [48].

### 2.3.1.2    Proactive Sessionization

This type of sessionization is based on the direct information delivered by a particular Web user. This might have fewer technical issues than reactive sessionization, but there are some details that will be discussed in what follows.

Besides user authentication, one of the basic elements considered for proactive sessionization are the Web browser cookies. These elements, despite their security issues as discussed in Section 2.2.1, can be used as a data repository which stores all end user activities from a Web browser. This client-side repository can be accessed and updated by code embedded in a Web page, and is usually considered by developers for persistence applications and improving the user experience when navigating the website. The information collected can be used to uniquely identify the user and track its session by sending back to the Web server only basic pieces of information to build a profile without comprising further privacy elements from the end user. However, there are no guarantees that more data will not be collected and used for applications without the user's consent.

### 2.3.1.3 Reactive Sessionization

When there is no information on which user has accessed a Web page, reactive sessionization methods have to be applied. The main components of a Web server log that are used in these type of methods are the IP address, the user-agent, the time when the Web page was accessed, and the URL requested before the Web page. Using simple rules based on IP and user-agent combinations, and considering their order based on the referral URLs and simple time constraints is a common practice, but does not consider the whole set of possibilities. For example, the Web user might use the browser's back button and break the order. If there is more than one user behind the IP address, it may not really be understood from whom the Web server logs are being generated.

However, despite the problems mentioned before, the most popular method for sessionization is based on setting rules on the maximum time that the user spends in a website [5]. After using the IP address and the user-agent for creating a partitioning over the Web server logs, a time window is used as threshold to exclude all records from a particular session. Given $t_0$ the first request timestamp for a session $s$, then a record with timestamp $t$ will be assigned to session $s$, if both IP address and user-agent are matched and if $t - t_0 \leq \theta$ [27]. A time window of $\theta = 30$ minutes is generally used to slice the records [40], and is based on experimental evidence that an average Web user spends that much time in a given website. Nowadays, this number seems to have a poor adjustment to what the current interaction between Web users and websites is, but has reported reasonable results [49].

Furthermore, the definition of a *page-stay-time-based* method has been proposed [4], where an additional time constraint $\delta$ is considered for each one of the records. Given $t_i, i \geq 0$ a record timestamp assigned to session $s$, a record with timestamp $t_j, j > i$ is assigned to $s$ if $t_j - t_i \leq \delta$. In this case, a threshold estimate of $\delta = 10$ minutes has been used as a conservative value which reportedly gives enough time to the Web user to load a Web page, and then review all its content.

An extra constraint that is not always considered because of the computational complexity overhead that adds to the sessionization process, is the referral condition where a record $i$ will only be added to the session $s$ if the IP address, the user-agent, the $\theta$ and $\delta$ time conditions, and the referrer for $i$ is already considered in $s$ [5, 32].

Further methods which consider semantic information for the sessions, ontology-based approaches have been developed to consider all previous constraints and the semantic information from the website that is being visited. This way, adding a semantic description to each URL, the algorithm can compute a semantic distance matrix $\Delta_{ij}$ to adjust the sessions which are not considering a minimum semantic threshold [20, 22].

### 2.3.1.4 Stochastic Sessionization Model

When dealing with applications involving the display of advertisements in search engine results, the click-through rate (*CTR*) from independent Web users is important to measure and thus determine the impact of the advertisement within the auction system that is being used. In this case, a stochastic approach can be used for modeling the user sessions [37].

Consider the separation of search engine results into "Web" (algorithmic), "Sponsored" (bidded), and "'Next' (from subsequent pages of search results). Given this, a user session is composed of the following events:

- Page request ($P$)
- Click in a webpage ($W$)
- Sponsored click ($O$)
- Next click to go through the multiple results of the search engine ($N$)
- Clicks that are not $W$, $O$, or $N$, denoted by $A$. (e.g. clicks on miscellaneous buttons like "Images" or "Videos" on Google search engine).

Then, the Event-Locality Pair (*ELP*) is defined as an ordered pair of (*event, page number*) where *event* $\in \{P, W, O, N, A\}$ and *page number* $\in \mathbb{N}$.

A user session $s$ can be modeled as a sequence of *ELP*s assuming that the next event in a session is impacted by previous events, the user session can be modeled as a Markov chain model. Let us represent the start of a session by a state '$S$'. The state space of the Markov chain can be defined as Equation 2.1, and the transition probability $\pi_{i,j}$ from state $i$ to $j$ can be estimated by Equation 2.2.

$$\text{Markov chain states} = \{S\} \cup \{\{P, W, N, O, A\} \times \mathbb{N}\} \tag{2.1}$$

$$\pi_{i,j} = \frac{Q_{i,j}}{Q_i} \tag{2.2}$$

where $Q_{i,j}$ is the number of instances where state $i$ is followed by state $j$ in the ELP sequences of all user sessions, and $Q_i = \sum_j Q_{i,j}$.

Based on this model, all transitions with high probability can be associated with normal behavior of users navigating the search engine results, while transitions with low probability can be classified as rare. If pricing strategies are related to cost-per-impression or cost-per-action, low probability or rare sessions could be classified as outliers and discarded from the set of actual sessions considered when computing the results of the bidding system.

After computing the complete transition matrix $\pi_{i,j}$ from click-stream data, each user session can be assigned by considering the likelihood score $\phi_k$ for user $k$, defined in Equation 2.3 from its sequence of ELP (*ELPs$^k$*) by multiplying the probabilities of its state transitions.

$$\phi_k = \prod_{i,j \in ELPs^k, i > j} \pi_{i,j} \tag{2.3}$$

This way, a user with a large ELP sequence is likely to get a small score, for which the Markovian Log-Likelihood for the ELP sequence of user $k$ ($MLH_{avg}^k$) is defined,

$$MLH_{avg}^k = \frac{\ln(\phi_k)}{|ELPs^k|} \qquad (2.4)$$

Where $|ELPs^k|$ is the total number of pairs considered in the ELP sequence of user $k$. This value can be considered as a measure of convergence between a user session and the normal (or popular usage). A higher value means that most of the ELP sequence transitions are normal, and a lower value means that the ELP sequence transitions are rare, and potentially outliers. However, this model might still not reflect the real interaction from the user with the search results, as a ELP sequence could have the same ELP repeated several times (which is rare) and might bias the results when there is another ELP sequence (which might be normal) with a lower frequency. In this case, the normal ELP sequence could be classified as atypical, and the rare ELP sequence as typical. For this, a multidimensional session model is proposed. Let, $P_t$ by the number of page requests, $W_t$ the number of clicks, $O_t$ the number of sponsored clicks, $N_t$ the number of next clicks, $A_t$ the number of any clicks, and $E = P_t + W_t + O_t + N_t + A_t$ the total number of events. Then, a user session $q$ can be represented as,

$$q = (MLH_{avg}^k, E, P_f, W_f, O_f, N_f, A_f) \qquad (2.5)$$

where $P_f = \frac{P_t}{E}$, $W_f = \frac{W_t}{E}$, $O_f = \frac{O_t}{E}$, $N_f = \frac{N_t}{E}$, and $A_f = \frac{A_t}{E}$. Based on this 7D vector, a user session outlier can be determined by approximating the probability measure around a 7D point $q$ by the Mahalanobis distance, defined as

$$d = \sqrt{(q-\mu)\Sigma^{-1}(q-\mu)^T} \qquad (2.6)$$

where $\mu$ represents the mean row vector and $\Sigma$ the covariance matrix for the complete dataset of click-stream 7D characteristics. When this distance is higher, then the density of the probability distribution around $q$ is lower, hence the rarity of $q$ is higher. To classify the outliers, the tail $x\%$ of the Mahalanobis distance distribution can be classified as atypical (it has been experimentally reported that this value can range from $x \in [0.5, 1]$ [37]).

### 2.3.1.5 Integer Programming Based Sessionization

Dell et al., in [6] presented a novel approach for Web user sessionization based on a Integer programming problem (Problem 2.7). Let $o$ be the order of a log registered during a session, and $|o|$ represent the maximum length of a session. Let $p, p'$ be Web pages, $r, r'$ Web log records, and $s$ a Web user session, where $r \in$ bpage$_r$ the set of entry logs that can be recorded immediately before record $r$ in the same session, and $r' \in$ first set of entry logs that must be first in a session.

Let $C_o$ be the objective function coefficient of having a record assigned to the $o^{th}$ position in a session, and $X_{os}$ be 1 if a log record $r$ is assigned to the $o^{th}$ position during session $s$, and 0 otherwise.

$$\max_{X_{ros}} \quad \sum_{ros} C_o X_{ros}$$

$$\text{s.t.} \quad \sum_{os} X_{ros} \le 1, \forall r$$

$$\sum_{r} X_{ros} \le 1, \forall o, s \tag{2.7}$$

$$X_{r,o+1,s} \le \sum_{r' \in \text{bpage}_r} X_{r',o,s}, \forall r, o, s$$

$$X_{ros} \in \{0, 1\}, \forall r, o, s$$

$$X_{ros} = 0, \forall r \in \text{first}, o > 1, s$$

The objective of this approach is to maximize the total number of records $r$ assigned to a particular session $s$ and position $o$, where a total reward of $\sum_{o' \le o} C_{o'}$ is obtained for any session of size $|o|$. As presented in [6], setting for example $C_3 = 1$ and $C_o = 0, \forall o \neq 3$ defines an objective function that maximizes the number of sessions of size three. After running several experiments with real data, researchers found that the expression for $C_o$ represented in Equation 2.8 has a better correlation coefficient and standard error when comparing the methodology with a set of real sessions. For more details, please refer to the work in [6].

$$C_o = \frac{3}{2} Log(o) + \frac{(o-3)^2}{12o} \tag{2.8}$$

Constraints defined in Problem 2.7 ensure that no record $r$ can be assigned to another session and position, each session has at most one record assigned for each ordered position, the proper ordering of records in the session are properly ordered, and there is only one first record $r \in$ first.

## 2.3.2    Dealing with Incomplete Information

Sometimes, client or proxy caching might cause missing references to objects that were accessed, but no records were kept in the Web server logs. The missing data might be hiding relevant information which would be interesting to consider when doing the Web usage analysis.

A simple heuristic to complete the missing information, is that given the structure of the website, would be to generate all potential candidate paths, and rank them by their number of "back" references. Candidate paths can be computed based on the actual entries in the Web logs, which could be used as a seed to generate all possible paths from the observed Web log entries. However, as the size of the site and the number of users increase in scale, the complexity of computing all potential paths and the most likely outcome is higher. For this, advanced heuristics considering the time spent in each page have been proposed [26], or path reconstruction algorithms considering as input the classification of entry and exit pages which could lead to a lower number of potential paths to match as best solution [40].

### 2.3.3 Model Assessment for Sessionization Algorithms

As previously discussed, the sessionization problem can be considered as a non-supervised classification problem where using some rules and heuristics, the algorithm has to decide whether a given set of pageviews corresponds to a particular session or not. However, the output of this process can be represented as a multi-classification problem where all pages retrieved are classified according to labels for a session identifier. For the evaluation of this output, it can be considered as a traditional multi-classification problem, where performance measures like *Precision*, *Recall* and the *F-measure* can be defined [44].

Let $C$ be the set of all relevant sessions that must be determined by a sessionization algorithm, $S$ the set of all sessions determined by the sessionization algorithm, and $M$ the matching sessions from the algorithm's output with the real sessions. Using these sets, the precision (the fraction of matched sessions that are relevant) and recall (the fraction of relevant sessions that are matched) can be defined as Equation 2.9 and Equation 2.10 respectively.

$$p = \frac{M}{S} \tag{2.9}$$

$$r = \frac{M}{C} \tag{2.10}$$

Furthermore, these measures can be combined in the *F-measure*, Equation 2.11, which represents the harmonic mean between *Precision* and *Recall*,

$$\text{F-measure} = \frac{2pr}{p+r} \tag{2.11}$$

The *F-measure* is a factor that yields the overall evaluation of both metrics, and can be used to decide whether the sessionization algorithm has acceptable results when this measure is higher [6]. However, as Web users change their behavior with respect to a particular website, these values must be evaluated over time to identify the potential concept-drift of the unsupervised classification problem [50].

Precision, recall and F-measure are the most basic metrics that can be computed to evaluate the quality of any sessionization heuristic. However, none of these metrics consider important characteristics like session containment, reconstruction, common fragments between predicted and real sessions, and the overlap. By using these concepts, the effectiveness of sessionization heuristics can be represented by more specific measures that could be fairer at the moment to compare results of these types of methods. For further details and explanation of the advanced sessionization quality metrics, please refer to Spiliopoulou in [40].

## 2.4 Profiling and Relevance Data Processing

A large component in processing Web usage data is related to the application of presenting the best possible information to the end user. Several examples of this can be identified from different applications and websites, such as item recommendation in Amazon[7], search engine results ranking in Google[8], or movie recommendation in Netflix[9]. To achieve this, the user has to declare the preference over a set of objects that a particular Web site can offer. These preferences have traditionally been considered explicit [39], implicit [21, 51] and mixed approaches. The explicit is what traditionally websites have already traditionally been doing, allowing each user to declare a global profile with preferences, or allowing the explicit rating of the information as is displayed in a regular user session over the website. Implicit preference declarations are mostly based on the click-through activity over the website, where the idea is that the type of content that the Web user consumes is the type of content that he would prefer to consume in future sessions.

As explicit rating and feedback is more reliable given that it shows the real preferences of the user, in some domains (e.g. text retrieval) it has been shown to be a cognitively demanding process [33]. This has been the main reason that researchers ad practitioners have been centered in improving the process of implicit ratings in several contexts, like videos [15, 45], Web searches [19, 52], and music [13], among other Web documents.

Mixed approaches have been shown to be complementary in certain domains like music recommendation systems [18, 49]. However, in domains like information search, mixed techniques have shown that implicit feedback is a marginal contribution when combined with explicit feedback [55].

When dealing with profiling, a single user might have several profiles which could be used depending on the particular intentions of the user at a given moment. For example, a user of a Web application that delivers music might want to have a profile for different interests related to the mood of the moment, but does not forget already defined profiles. The multi-profiling problem can be addressed by asking the user to select the profile he wants to be classified at when accessing the application, but depending on the business model, the multi-profiling strategy could be something that the practitioner would like to explore. Recently, approaches considering the inclusion of the Ostensive Model for information needs [2] have been incorporated when dealing with multiple-profiling problems [14].

In the following, traditional relevance feedback approaches with explicit ratings are presented in Section 2.4.1, and a brief introduction to the main processing challenges of eye tracking methodologies are discussed in Section 2.4.2.

---

[7] www.amazon.com [last accessed April 18, 2012].

[8] www.google.com [last accessed April 18, 2012].

[9] www.netflix.com [last accessed April 18, 2012].

## 2.4.1 Relevance Feedback and Explicit Ratings

To discuss the details that surround the relevance feedback from explicit rating and how to include that in a particular application, let's think about the use case where a particular end user is looking for a set of objects given an input query $q$, and the feedback is explicitly returned for the results obtained. In this context, the Rocchio algorithm for relevance feedback is one of the most used methods. However, to deal with uncertainty and and static values which could dynamically change the relevance and its parameters' likelihood, further approaches based on probabilistic methods have been proposed. In the following, both the Rocchio algorithm and a probabilistic approach will be reviewed.

### 2.4.1.1 Rocchio Algorithm for Relevance Feedback

The Rocchio algorithm [36] is one of the first approaches to consider the relevance feedback in an information retrieval (IR) context. This method can nowadays can be closely related to Web usage Mining as the information is shared within a system by the explicit feedback from a Web user, which is used to feed back the algorithm that computes a given output. In the original case, the input and output representation of the feedback is determined by the vector space model (VSM) proposed by Salton [38].

The basic concept underlying this method is to find the optimal vector $\mathbf{q}_{opt}$ that solves the maximum similarity between relevant objects while minimizing the similarity for non-relevant objects.

Let $D_r$ be the set of relevant objects (declared explicitly by the user) among those retrieved by an IR model, $D_n$ the set of non-relevant objects (declared explicitly by the user) retrieved by a given IR model, $C_r$ the set of relevant objects among all objects in the collection (declared explicitly by the all the users), and $N$ the total number of objects in the collection. Then the optimal query $\mathbf{q}_{opt}$ for a particular query can be determined by the following equation,

$$\mathbf{q}_{opt} = \frac{1}{|C_r|} \sum_{\forall \mathbf{d}_j \in C_r} \mathbf{d}_j - \frac{1}{N - |C_r|} \sum_{\forall \mathbf{d}_j \notin C_r} \mathbf{d}_j \qquad (2.12)$$

where $|C_r|$ is the cardinality of the set $C_r$, $\mathbf{d}_j$ is a weighted vector associated with retrieved object $d_j$. One of the main problems with this methodology is that the set $C_r$ is not given in advance and a cold start problem needs to be managed directly with the end users. For this, the Roccio algorithm [36] considers an initial query $\mathbf{q}_0$ which is combined with terms originated from a partial knowledge of the relevance of the objects to be retrieved. In this way, a modified query $\mathbf{q}_m$ is computed following Equation 2.12,

$$\mathbf{q}_m = \alpha \cdot \mathbf{q}_0 + f(\beta) \sum_{\forall \mathbf{d}_j \in D_r} \mathbf{d}_j + g(\gamma, D_n) \qquad (2.13)$$

where $\alpha, \beta, \gamma$ are tuning parameters, function $f$ could be the identity function where $f(\beta) = \beta$ or $f(\beta) = \frac{\beta}{N_r}$ (for the standard Rocchio representation), and several representations for the function $g$ have been proposed. Some of them are presented in Equation 2.14.

$$g(\gamma, D_n) = \begin{cases} \frac{\gamma}{N_n} \sum_{\forall \mathbf{d}_j \in D_n} \mathbf{d}_j & \text{for standard Rocchio model [36]} \\ \gamma \sum_{\forall \mathbf{d}_j \in D_n} \mathbf{d}_j & \text{for Ide regular model [16]} \\ \gamma \cdot \text{max\_rank}(D_n) & \text{for Ide "Dec Hi" model [16]} \end{cases} \quad (2.14)$$

where max_rank($D_n$) is a method that returns the highest ranked non-relevant object. Further tuning on parameters $\alpha$, $\beta$, and $\gamma$ can be performed. For example, based on the idea that the information related to the relevant objects is more important that the information contained in the non-relevant objects, the $\beta$ parameter should be larger than $\gamma$. The $\alpha$ parameter, as is related directly to the original IR query, might have some important information to determine which objects are relevant so its value is generally set to $\alpha = 1$.

### 2.4.1.2 Probabilistic Approach for Relevance Feedback

This model ranks objects similar to a given IR query $q$ based on a different approach than that presented in Section 2.4.1.1. As presented in [1], let $R$ be the set of objects known to be relevant and $\overline{R}$ the set of non-relevant objects. Let $P(R|\mathbf{d}_j, q)$ be the probability that object $d_j$ represented by $\mathbf{d}_j$ is relevant to query $q$. Then, the similarity between an object $\mathbf{d}_j$ and its query $q$ can be defined as,

$$sim(\mathbf{d}_j, q) = \frac{P(R|\mathbf{d}_j, q)}{P(\overline{R}|\mathbf{d}_j, q)} \quad (2.15)$$

This similarity can be used by both practitioners and researchers to rank all relevant objects for a given input query $q$, and return them accordingly to the end users. Following the same idea than the methodology described in Section 2.4.1.1, this method is closely related to the processing of Web usage data that is input by an end user, as is generated by the explicit usage of an IR system which is collecting at all times relevant information to update the following versions of the model.

Based on several assumptions[10], it can be shown that,

$$sim(\mathbf{d}_j, q) = \sum_{k_i \in q \wedge k_i \in d_j} \log\left(\frac{p_{iR}}{1 - p_{iR}}\right) + \log\left(\frac{1 - q_{iR}}{q_{iR}}\right) \quad (2.16)$$

where $p_{iR} = P(k_i|R, q)$ is the probability that a feature $k_i$ is present in object selected randomly from the set of relevant objects $R$, and $q_{iR} = P(k_i|\overline{R}, q)$ is the probability that a feature $k_i$ is present in an object selected randomly from the set of non-relevant objects $\overline{R}$. As might be anticipated by some readers, this definition is originally

---

[10] For more details, please refer to [1], pg. 79.

determined for the case where objects are documents and $k_i$ are terms within the document, but it can be extended to a more general case when considering that input variables are independent from each other.

When the algorithm does not have enough information to estimate its probabilities, in time $t = 0$, the $p_{iR}$ can be defined as 0.5 and $q_{iR} = \frac{n_i}{N}$, where $n_i$ is the number of objects with feature $k_i$ and $N$ is the total number of objects. Considering previous values, the expression for the similarity is,

$$sim(\mathbf{d}_j, q) = \sum_{k_i \in q \wedge k_i \in d_j} \log\left(\frac{N - n_i}{n_i}\right) \tag{2.17}$$

Then, as data is collected over time, the probabilities can be updated according to Equation 2.18 and Equation 2.19

$$P(k_i | R, q) = \frac{n_{r_i}}{N_r} \tag{2.18}$$

$$P(k_i | \overline{R}, q) = \frac{n_i - n_{r_i}}{N - N_r} \tag{2.19}$$

where $n_{r_i}$ is the total number of objects in $D_r$ containing $k_i$ feature, $D_r$ (recalling Section 2.4.1.1) is the set of relevant objects (declared explicitly by the user) among those retrieved by an IR model, and $N_r$ is the total number of objects in $D_r$. To avoid problems related to small numbers when evaluating the probabilities, small adjustment factors have been introduced. Accordingly, updating previous equations, the final expressions are represented by Equation 2.20 and Equation 2.21,

$$P(k_i | R, q) = \frac{n_{r_i} + 0.5}{N_r + 1} \tag{2.20}$$

$$P(k_i | \overline{R}, q) = \frac{n_i - n_{r_i} + 0.5}{N - N_r + 1} \tag{2.21}$$

## 2.4.2  Eye Tracking Data Processing

Using today's technology, the eye of a Web user can be tracked to perform advanced experiments for the evaluation of the Web usage and the navigation of a given website. This process, which currently is performed in small labs with a limited number of end users, has provided enough information to explain which are the spots where in average humans will focus when navigating a particular website, and subsequently which spots could be used to communicate correctly to the Web user and try to maximize the conversion rate with respect to the objective of the website.

According to Rayner [35], ocular behavior can be classified into four types: fixation, saccades, pupil dilatation, and scan paths. Among these, fixation is the most relevant when using eye tracking devices to understand where humans focus at the moment of navigating through a website. This type of behavior lasts between 200 – 300 milliseconds, which makes it easier to accurately track with a device. All other ocular behavior types last for a shorter period of time, and do not provide enough time for the suspect to focus on the information that is being presented.

Some of the traditional eye tracking techniques are based on scleral lens analysis [9], the evaluation of photo and video oculography [3, 9], and the reflex of the cornea and the center of the pupil based on video analysis [34]. The latter approach is among the most used for current eye tracking research. This method consists of an infrared camera located under the computer screen, with a image-processing software to locate and identify the reflex of the cornea and the pupil center. Based on these characteristics, it is possible to determine the ocular movements which is used to compute the attention center of the end users [34]. When the system is activated, an infrared light from an LED is directed towards the end user to create special reflexes making the eye movement easy to track. The light enters into the retina and a great part is reflected, giving the pupil the shape of a well-defined bright disc. Once the data is being tracked by the software, a vector is recorded with several parameters measured. In general terms, the data generated by an eye-tracker device can be summarized by the following:

- Timestamp: Date and time in *milliseconds* on when the data was collected.
- Gaze Point $X$ Left: Horizontal position in the screen that the left eye is observing.
- Gaze Point $Y$ Left: Vertical position in the screen that the left eye is observing.
- Cam $X$ Left: The horizontal location of the left pupil in the camera image.
- Cam $Y$ Left: The vertical location of the left pupil in the camera image.
- Distance Left: Distance in *millimeters* from the eye tracker to the left eye.
- Pupil Left: Diameter of the left eye in *millimeters*.
- Gaze Point $X$ Right: Horizontal position in the screen that the right eye is observing.
- Gaze Point $Y$ Right: Vertical position in the screen that the right eye is observing.
- Cam $X$ Right: The horizontal location of the right pupil in the camera image.
- Cam $Y$ Right: The vertical location of the right pupil in the camera image.
- Distance Right: Distance in *millimeters* from the eye tracker to the right eye.
- Pupil Right: Diameter of the right eye in *millimeters*.

The previously mentioned eye-tracker data collection strategy is to be adjusted to each person given their own characteristics, where the device's calibration process shows a point to the user which has to be observed within a margin of error and for a minimum amount of time. Then, the device determines the center of the pupil and the reflex of the cornea in a specific location over the monitor. This process is repeated several times showing other points in the monitor to acquire the best possible precision. However, all measurements might change according to the physical properties of the test subject. Also, a validation process records information for each eye independently, and when processing eye tracking data it is important to consider these types of records. For example, the device might register that the left eye was focused in a specific area of the monitor, while the right eye is not being tracked by the same event. This way, the left eye record can be considered as an outlier and removed from the data to be analyzed.

The data generated by eye tracking devices is hard to process and several algorithms have been proposed by researchers to improve the definition of which are the paths that the eye was declaring as relevant while the subject was navigating a given website. According to [25], these algorithms are Velocity Threshold Identification (I-VT), Dispersion Threshold Identification (I-DT), Minimum Spanning Tree Identification (MST), Hidden Markov Model Identification (I-HMM), and Kalman Filter Identification (I-KF). Each one of these methods presents different alternatives to approximate the movement of the eyes. A full review of these classification methods is presented in [24].

The tracking of eye movement might be interesting, but without a clear picture of what the user understands from the navigation, all sorts of conclusions could be reached. One approach that tackles the understanding of the eye movement by pairing it with contextual understanding, is presented by Nakayama & Hayashi [31]. In this approach, the collection of 12 directional features of eye movement across a 2-D space, while observing correct and incorrect responses of question statements is proposed. In this experiment, by using machine learning algorithms (particularly Support Vector Machines), the process of classifying whether a sentence statement is correct or incorrect was successfully predicted by using the tracked features.

Further methods related to the evaluation of eye movement when mixing both implicit and explicit relevance feedback have been proposed. For example in the case of implicit feedback, Faro et al. in [10] evaluated both eye tracking data together with implicit information of the user in multimedia retrieval problems, where the quality of results are evaluated in the context of content-based image retrieval. In the case of explicit relevance feedback, in a context similar to the evaluation of content-based image retrieval systems, as proposed by [54], eye tracking data is successfully merged together with relevance feedback data.

As mentioned before, in terms of Web usage mining, the information collected by eye tracking devices has been used to improve the understanding of how end users' attention is distributed over a particular website. Granka et al. [11] concluded based on the Web user behavior data collected using eye tracking devices, that the first two links of a search engine are commonly selected by Web users. Also, in this paper, researchers studied the size of elements and the density of information within a Web page, and how they influence the capacity of driving the end user's attention. They concluded that there is no significant contribution when the user starts directing his attention to any of the elements. Furthermore, they discovered that most end users have developed the capacity to ignore the areas where the advertisement is displayed in Web pages. These numbers (published in the year 2004) might not reflect what is happening today, as the behavior might have changed based on the technological advances and the wide adoption and penetration of the Internet in our current society.

## 2.5  A Discussion on Privacy in the World Wide Web

Whenever data collection processes include the tracking of personal information to determine a particular profile for a unique user, it involves deep privacy issues. The identification of personal data in some contexts and using some methodologies can be illegal depending on the regulations defined by each country. Furthermore, the privacy implications associated with the data collection systems have brought a wide discussion which relates legal issues, social norms, market dynamics and economic impact. Each one of these concepts involves both technical and ethical implications that have to be addressed by both practitioners and researchers when building systems that track sensitive information from Web users.

As discussed by Moloney & Bannister in [30], privacy can be related to the following concepts:

1. The control of information concerning an individual.
2. The right to prevent commercial publicity of one's own name and image.
3. The right to make certain personal and intimate decisions free from government interference.
4. To be free from physical invasion of one's home or person.

Each one of the previous concepts might be compromised when using invasive tracking strategies such as Web browser cookies [53]. Whenever a Web usage and profiling methodology is developed, beyond a discussion of its technical capabilities, it should be tested in terms of legal, social and ethical norms.

Recently, in late 2011 and early 2012, in the United States (U.S.) Congress a heated legislative debate on the way the information tracked by third parties (e.g. Internet Service Providers, Web sites, etc.) from Web users was going to be used took place. These laws were referred to as the Stop Online Piracy Act (SOPA)[11], the PROTECT IP Act (PIPA)[12], and the Cyber Intelligence Sharing and Protection Act (CISPA)[13]. They have been extensively discussed by several parties involved in internet activities, from large companies to Web users, and this discussion has reached the attention of thousands of internet users beyond the U.S. The main issues related to these laws are the lack of privacy and the actions that a centralized organism could take without any notifications to the internet users involved in piracy allegations, that could be placed by any one at any time.

One of the main privacy breaches that many users are not considering when accepting the terms and conditions of a particular website, is that related to the

---

[11] For more information, please refer to
`http://thomas.loc.gov/cgi-bin/bdquery/z?d112:HR03261:` [last accessed April 18, 2012].

[12] For more information, please refer to `http://www.leahy.senate.gov/imo/media/doc/BillText-PROTECTIPAct.pdf` [last accessed April 18, 2012].

[13] For more information, please refer to
`http://thomas.loc.gov/cgi-bin/query/z?c112:H.R.3523:` [last accessed April 18, 2012].

information that they are explicitly and/or implicitly sharing with a certain party (which could be a second or third party). There is an infamous quote by Andrew Lewis[14] that describes what is happening without the proper education of the end users, "If you are not paying for it, you're not the customer; you're the product being sold".

An example of the latter is related to the interactions that billions of users have with the well-known online social networks. Whenever an end user shares a message, picture, or a click over these sites, he or she is giving all the rights to the owner of the data servers to use or share that information without any notification to the end user. Each time the end user opts in to a social network by creating an account, is explicitly accepting the terms and conditions. Furthermore, when the user opts out of a social network, all the information is still property of the server owners, and there is no law or mechanism that could be used to retrieve and/or delete all personal information from the website.

Explicit information tracking strategies are present in several contexts, and other domains, for example mobile devices like iPhone, which in late 2011, was discovered that Apple, Inc. was using the information related to nearby mobile networks to collect personal information of every active device. This tracking issues are not explicitly shared with the end users, but in terms of the disclosure terms of the company, are included within a legal framework.

All these issues are not illegal, but certainly are not correctly transmitted to the end users, which many times are not even allowed to access the complete records of their information. Some companies have given the end users the possibility of downloading the complete information traces that a first party has obtained from the activities of the user within a respective website, like Google Takeout[15]. Another example is the OpenPaths project[16] which the New York Times has released for accessing the personal information that is being tracked from each iPhone device, giving the possibility of sharing the data with researchers and practitioners with the explicit permissions of the owner of the device.

## 2.6 Summary

As we navigate the Web, our footsteps and fingerprints are collected by several Web servers around the world. This information, basically represented by Web server logs, gives the potential of understanding specific details of the behavior and usage within a given website. These knowledge-generation techniques are known within the Web usage mining framework, and are a strong component when considering whether to leverage the available data to improve the business of a given project or company based on Web activities.

---

[14] http://www.metafilter.com/user/15556 [last accessed April 20, 2012].

[15] www.google.com/takeout [last accessed April 18, 2012].

[16] openpaths.cc [last accessed April 18, 2012].

In Section 2.2, fundamentals about the collection and processing of Web usage data are presented. First, there is a discussion of the data cleaning and merging process to consider when dealing with raw Web server logs. The cleaning process is closely related to the removal of non-relevant records associated with Web crawlers and objects embedded within a Web page. The merging process is usually related to the fact that information might be distributed over several Web servers, and a synchronization process must be performed.

Once the data is cleaned and merged, the process of user session identification (or sessionization) is fundamental for most of Web usage mining applications. For this, several methods have been proposed, ranging from basic heuristics to complex optimization formulations. Some of the most interesting techniques and their respective definitions are presented in Section 2.3.

Applications regarding how to collect and process information about the feedback of Web users is presented. In this case, the nature of the data can be either explicit or implicit, contributed by the end users as they interact with a particular website. In Section 2.4, main concepts regarding implicit and explicit relevance feedback collection and processing techniques are introduced, as well as a discussion on how eye tracking devices have been used to improve the understanding of the end user.

Finally, in Section 2.5 a discussion on privacy and the legal issues surrounding this type of application is presented. In this case, there is an extended discussion on how and what information has been collected by well-known websites and a quick review of the issues related to how the future of privacy over the World Wide Web might be compromised by legislative initiatives from the United States Congress, which have been publicly discussed all over the world.

**Acknowledgements.** This work has been partially supported by the Chilean Millennium Institute of Complex Engineering Systems (ICM: P-05-004-F, CONICYT: FBO16).

# References

1. Baeza-Yates, R., Ribeiro-Neto, B.: Modern Information Retrieval, 2nd edn. Addison-Wesley Publishing Company, USA (2008)
2. Campbell, I., Van Rijsbergen, C.J.: The ostensive model of developing information needs, pp. 251–268. The Royal School of Librarianship (1996)
3. Cerrolaza, J.J., Villanueva, A., Cabeza, R.: Taxonomic study of polynomial regressions applied to the calibration of video-oculographic systems. In: Proceedings of the 2008 Symposium on Eye Tracking Research &#38; Applications, ETRA 2008, pp. 259–266. ACM, New York (2008)
4. Cooley, R., Mobasher, B., Srivastava, J.: Data preparation for mining world wide web browsing patterns. Knowl. Inf. Syst. 1(1), 5–32 (1999)
5. Das, R., Turkoglu, I.: Creating meaningful data from web logs for improving the impressiveness of a website by using path analysis method. Expert Syst. Appl. 36(3), 6635–6644 (2009)

6. Dell, R.F., Román, P.E., Velásquez, J.D.: Web user session reconstruction using integer programming. In: Proceedings of the 2008 IEEE/WIC/ACM International Conference on Web Intelligence and Intelligent Agent Technology - Volume 01, WI-IAT 2008, pp. 385–388. IEEE Computer Society, Washington, DC (2008)
7. Demir, G.N., Goksedef, M., Etaner-Uyar, A.S.: Effects of session representation models on the performance of web recommender systems. In: Proceedings of the 2007 IEEE 23rd International Conference on Data Engineering Workshop, ICDEW 2007, pp. 931–936. IEEE Computer Society, Washington, DC (2007)
8. Doran, D., Gokhale, S.S.: Web robot detection techniques: overview and limitations. Data Min. Knowl. Discov. 22(1-2), 183–210 (2011)
9. Duchowski, A.T.: Eye Tracking Methodology: Theory and Practice. Springer-Verlag New York, Inc., Secaucus (2007)
10. Faro, A., Giordano, D., Pino, C., Spampinato, C.: Visual attention for implicit relevance feedback in a content based image retrieval. In: Proceedings of the 2010 Symposium on Eye-Tracking Research & Applications, ETRA 2010, pp. 73–76. ACM, New York (2010)
11. Granka, L.A., Joachims, T., Gay, G.: Eye-tracking analysis of user behavior in www search. In: Proceedings of the 27th Annual International ACM SIGIR Conference on Research and Development in Information Retrieval, SIGIR 2004, pp. 478–479. ACM, New York (2004)
12. Gündüz, Ş., Tamer Özsu, M.: A web page prediction model based on click-stream tree representation of user behavior. In: Proceedings of the Ninth ACM SIGKDD International Conference on Knowledge Discovery and Data Mining, KDD 2003, pp. 535–540. ACM, New York (2003)
13. Hoashi, K., Matsumoto, K., Inoue, N.: Personalization of user profiles for content-based music retrieval based on relevance feedback. In: Proceedings of the Eleventh ACM International Conference on Multimedia, MULTIMEDIA 2003, pp. 110–119. ACM, New York (2003)
14. Hopfgartner, F., Hannah, D., Gildea, N., Jose, J.M.: Capturing multiple interests in news video retrieval by incorporating the ostensive model. In: PersDB 2008, 2nd International Workshop on Personalized Access, Profile Management, and Context Awareness: Databases, Electronic Proceedings, pp. 48–55 (2008)
15. Hopfgartner, F., Jose, J.: Evaluating the implicit feedback models for adaptive video retrieval. In: Proceedings of the International Workshop on Workshop on Multimedia Information Retrieval, MIR 2007, pp. 323–331. ACM, New York (2007)
16. Ide, E.: New Experiments in Relevance Feedback. Prentice-Hall, Englewood Cliffs (1971)
17. Ivancsy, R., Juhasz, S.: Analysis of web user identification methods. World Academy of Science, Engineering, and Technology 34, 34–59 (2007)
18. Jawaheer, G., Szomszor, M., Kostkova, P.: Comparison of implicit and explicit feedback from an online music recommendation service. In: Proceedings of the 1st International Workshop on Information Heterogeneity and Fusion in Recommender Systems, HetRec 2010, pp. 47–51. ACM, New York (2010)
19. Joachims, T., Granka, L., Pan, B., Hembrooke, H., Radlinski, F., Gay, G.: Evaluating the accuracy of implicit feedback from clicks and query reformulations in web search. ACM Trans. Inf. Syst. 25(2) (April 2007)
20. Jung, J.J., Jo, G.-S.: Semantic outlier analysis for sessionizing web logs. In: Proceedings of the 1st European Web Mining Forum, EWMF 2003, Croatia (2003)
21. Kelly, D., Teevan, J.: Implicit feedback for inferring user preference: a bibliography. SIGIR Forum 37(2) (September 2003)

22. Khasawneh, N., Chan, C.-C.: Active user-based and ontology-based web log data pre-processing for web usage mining. In: Proceedings of the 2006 IEEE/WIC/ACM International Conference on Web Intelligence, WI 2006, pp. 325–328. IEEE Computer Society, Washington, DC (2006)
23. Kohonen, T.: Self-organizing maps. Springer-Verlag New York, Inc., Secaucus (1997)
24. Komogortsev, O.V., Jayarathna, S., Koh, D.H., Gowda, S.M.: Qualitative and quantitative scoring and evaluation of the eye movement classification algorithms. Technical Reports-Computer Science, San Marcos, Texas, Texas State University (2009)
25. Komogortsev, O.V., Jayarathna, S., Koh, D.H., Gowda, S.M.: Qualitative and quantitative scoring and evaluation of the eye movement classification algorithms. In: Proceedings of the 2010 Symposium on Eye-Tracking Research & Applications, ETRA 2010, pp. 65–68. ACM, New York (2010)
26. Li, Y., Feng, B., Mao, Q.: Research on path completion technique in web usage mining. In: Proceedings of the 2008 International Symposium on Computer Science and Computational Technology - Volume 01, ISCSCT 2008, pp. 554–559. IEEE Computer Society, Washington, DC (2008)
27. Liu, B.: Web Data Mining: Exploring Hyperlinks, Contents, and Usage Data (Data-Centric Systems and Applications). Springer-Verlag New York, Inc., Secaucus (2006)
28. Mobasher, B., Dai, H., Luo, T., Nakagawa, M.: Effective personalization based on association rule discovery from web usage data. In: Proceedings of the 3rd International Workshop on Web Information and Data Management, WIDM 2001, pp. 9–15. ACM, New York (2001)
29. Mobasher, B., Dai, H., Luo, T., Nakagawa, M.: Discovery and evaluation of aggregate usage profiles for web personalization. Data Min. Knowl. Discov. 6(1), 61–82 (2002)
30. Moloney, M., Bannister, F.: A privacy control theory for online environments. In: Proceedings of the 42nd Hawaii International Conference on System Sciences, HICSS 2009, pp. 1–10. IEEE Computer Society, Washington, DC (2009)
31. Nakayama, M., Hayashi, Y.: Estimation of viewer's response for contextual understanding of tasks using features of eye-movements. In: Proceedings of the 2010 Symposium on Eye-Tracking Research & Applications, ETRA 2010, pp. 53–56. ACM, New York (2010)
32. Nasraoui, O., Soliman, M., Saka, E., Badia, A., Germain, R.: A web usage mining framework for mining evolving user profiles in dynamic web sites. IEEE Trans. on Knowl. and Data Eng. 20(2), 202–215 (2008)
33. Nichols, D.M.: Implicit rating and filtering. In: In Proceedings of the Fifth DELOS Workshop on Filtering and Collaborative Filtering, pp. 31–36 (1998)
34. Poole, A., Ball, L.J.: Eye Tracking in Human-Computer Interaction and Usability Research: Current Status and Future (2005)
35. Rayner, K.: Eye movements in reading and information processing: 20 years of research. Psychological Bulletin 124(3), 372–422 (1998)
36. Rocchio, J.J.: Relevance Feedback in Information Retrieval. Prentice-Hall, Englewood Cliffs (1971)
37. Sadagopan, N., Li, J.: Characterizing typical and atypical user sessions in clickstreams. In: Proceedings of the 17th International Conference on World Wide Web, WWW 2008, pp. 885–894. ACM, New York (2008)
38. Salton, G., Lesk, M.E.: Computer evaluation of indexing and text processing. J. ACM 15(1), 8–36 (1968)
39. Salton, G., Buckley, C.: Improving retrieval performance by relevance feedback. In: Readings in Information Retrieval, pp. 355–364. Morgan Kaufmann Publishers Inc., San Francisco (1997)

40. Spiliopoulou, M., Mobasher, B., Berendt, B., Nakagawa, M.: A framework for the evaluation of session reconstruction heuristics in web-usage analysis. INFORMS Journal on Computing 15(2), 171–190 (2003)
41. Stevanovic, D., An, A., Vlajic, N.: Detecting Web Crawlers from Web Server Access Logs with Data Mining Classifiers. In: Kryszkiewicz, M., Rybinski, H., Skowron, A., Raś, Z.W. (eds.) ISMIS 2011. LNCS, vol. 6804, pp. 483–489. Springer, Heidelberg (2011)
42. Suda, B.: Using microformats, 1st edn. O'Reilly (2006)
43. Tan, P.-N., Kumar, V.: Discovery of web robot sessions based on their navigational patterns. Data Min. Knowl. Discov. 6(1), 9–35 (2002)
44. Tsoumakas, G., Katakis, I.: Multi-label classification: An overview. International Journal of Data Warehousing and Mining 3(3), 113 (2007)
45. Vallet, D., Hopfgartner, F., Jose, J.M., Castells, P.: Effects of usage-based feedback on video retrieval: A simulation-based study. ACM Trans. Inf. Syst. 29(2), 11:1–11:32 (2011)
46. Velasquez, J.D., Palade, V.: Building a knowledge base for implementing a web-based computerized recommendation system. International Journal of Artificial Intelligence Tools 16(5), 793–828 (2007)
47. Velasquez, J.D., Palade, V.: A knowledge base for the maintenance of knowledge extracted from web data. Knowledge?Based Systems Journal 20(3), 238–248 (2007)
48. Velásquez, J.D., Dujovne, L.E., L'Huillier, G.: Extracting significant website key objects: A semantic web mining approach. Eng. Appl. Artif. Intell. 24(8), 1532–1541 (2011)
49. Velásquez, J.D., Palade, V.: Adaptive Web Sites: A Knowledge Extraction from Web Data Approach. IOS Press, Amsterdam (2008)
50. Wang, S., Schlobach, S., Klein, M.: What Is Concept Drift and How to Measure It? In: Cimiano, P., Pinto, H.S. (eds.) EKAW 2010. LNCS, vol. 6317, pp. 241–256. Springer, Heidelberg (2010)
51. White, R.W., Ruthven, I., Jose, J.M.: The Use of Implicit Evidence for Relevance Feedback in Web Retrieval. In: Crestani, F., Girolami, M., van Rijsbergen, C.J.K. (eds.) ECIR 2002. LNCS, vol. 2291, pp. 93–109. Springer, Heidelberg (2002)
52. White, R.W., Ruthven, I., Jose, J.M., Van Rijsbergen, C.J.: Evaluating implicit feedback models using searcher simulations. ACM Trans. Inf. Syst. 23(3) (July 2005)
53. Yue, C., Xie, M., Wang, H.: Automatic cookie usage setting with cookiepicker. In: Proceedings of the 37th Annual IEEE/IFIP International Conference on Dependable Systems and Networks, DSN 2007, pp. 460–470. IEEE Computer Society, Washington, DC (2007)
54. Zhang, Y., Fu, H., Liang, Z., Chi, Z., Feng, D.: Eye movement as an interaction mechanism for relevance feedback in a content-based image retrieval system. In: Proceedings of the 2010 Symposium on Eye-Tracking Research & Applications, ETRA 2010, pp. 37–40. ACM, New York (2010)
55. Zigoris, P., Zhang, Y.: Bayesian adaptive user profiling with explicit & implicit feedback. In: Proceedings of the 15th ACM International Conference on Information and Knowledge Management, CIKM 2006, pp. 397–404. ACM, New York (2006)

# Chapter 3
# Cognitive Science for Web Usage Analysis

Pablo E. Román and Juan D. Velásquez

**Abstract.** Web usage mining is the process of extracting patterns from web user's preferences and browsing behavior. Furthermore, the web user behavior refers to the user's activities in a web site. Cognitive science is a multi-disciplinary approach used for the understanding of human behavior, whose aims is to develop models of information processing in the real brain. Therefore, cognitive sciences can have direct application to web usage mining. In this chapter, some state-of-the-art psychology theories are presented in the context of web usage analysis. In spite of the complexity of neural processes in the brain, stochastic models based on diffusion can be used to explain a decision-making process, and this has been experimentally tested. Diffusion models and theirs application to describe web usage are reviewed in this chapter. An example of application of cognitive science to web usage mining is also presented.

## 3.1 Introduction

The Web has recently been transformed into a highly competitive and lucrative market. One of the highest-earning companies today is Google™ at several billion dollars per year [33]. The common factor among these companies, including Facebook™ and Amazon™, is having a thorough knowledge of the preferences of their users. We can say that the behavioral models available to these companies are partly responsible for their economic success. Cognitive science has developed theories of mind based on complex representations and computational procedures. Research

Pablo E. Román · Juan D. Velásquez
Web Intelligence Consortium Chile Research Centre, Department of Industrial
Engineering School of Engineering and Science, University of Chile,
Av. República 701, Santiago, Chile, P.C. 837-0720
e-mail: proman@ing.uchile.cl, jvelasqu@dii.uchile.cl

J.D. Velásquez et al. (Eds.): Advanced Techniques in Web Intelligence-2, SCI 452, pp. 35–73.

perspectives on the application of such theories to web research are listed in this chapter.

Psychology has been studying the process of decision making since ancient Greek times ("perception" for Protagoras). Modern psychology defines *cognition* as the mental faculty of processing information, applying knowledge and changing preferences. In this sense, perception is the cognitive process of attaining awareness or of understanding sensory information. More recently, economists have applied such concepts to explain consumers' behavior. Both disciplines focus on their own specific goals, yet nowadays application from one field into another has produced fruitful results and resulted in several Nobel prizes (e.g. [42]). In this chapter a review of current psychological models of the decision-making process are presented.

Web usage mining is the analysis of web users' browsing activity by means of machine learning techniques. Accordingly, web usage patterns can be used for analyzing web user preferences. The process of knowledge discovery in databases (KDD) [9] consists of data pre-processing (filtering and transformation), data mining, and interpretation and evaluation. Once data have been consolidated (WebWarehouse) [95], specific machine learning algorithms are applied in order to extract patterns regarding the usage of the web site. As a result of this process several applications can be implemented and integrated on adaptive web sites, such as recommender systems and revenue management marketing, among others. However, the problems connected to the customization of web sites that are geared to improving sales are somewhat challenging.

Recently, the one million-dollar NetFlix™ prize [53] was contested for three years without a winner. NetFlix™is an on-line DVD rental company driven by online movie recommendations, based on its customers' movie ratings. The contest required to improve the company's forecasting algorithm for each user ratings. The 2009 winning team used a collaborative filtering algorithm for recommendations [88], improving the performance of the NetFlix predictive algorithm by nearly 10%. This is an example of how difficult is to describe web user behavior, but also it highlights the real-life importance of such predictions. Notwithstanding the prize and the three-year window, only a modest improvement in performance was achieved, in spite of years of data mining research. The aim of this chapter is to explore unorthodox research streams based on cognitive sciences in order to illuminate new research directions.

Psychology has been engaged in several paradigms which focus on describing the mind's choice-selection process. Roughly two categories can be identified, perceptual choice-based and motivational valued-based choice selection. The first one relates to a large series of behavioral and neurophysiological experiments. Behavioral experiments analyze decision making and reaction time. Physiology experiments take into account the time evolution of neural tissue electric potential firing rates before reaching a decision. This last point of view is related to a mesoscopic dynamical description of natural science phenomena. A value-based choice selection approach has traditionally been the basis of many economic theories. This is the utilitarian scheme that represents a subject's preferences by a function dependent on the goods characteristic called "utility". In this theory, subjects tend to select the alternative with maximal utility. Nevertheless, this point of view does not consider the underlying characteristics of the process of decision making in isolation, rather it can be considered more as a static description.

Furthermore, two currents separate cognitive science, symbolicism and connectionism. The symbolic point of view considers that the mind works as a symbol manipulator and processor within an algorithmic-level framework. Computer chess games are examples of a further application of symbolicism. The connectionism point of view considers that cognition can be modeled as the processing of interconnected neural networks. The first approach corresponds to a higher level of analysis, where artificial intelligence and economics have their foundations, recovering empirical phenomena but not physical ones. The second approach relates to mimicking physical phenomena using simpler abstractions based on first principles, but not without difficulties in explaining higher-level mind results. On the symbolic side, approaches from utilitarian theory to games theory can be found, which are known as "preferential choices."

However, a degree of controversy exists between researchers of both sides in defining a unified point of view for describing the decision-making process [27]. Nevertheless, recent advances in computational neuroscience suggest a promising new research direction, despite symbolism having already reached a mature level.

The rest of this chapter is organized as follows. Section 3.2 provides a brief summary of preferential choice models. Such theories have been applied to model different behaviors on the web. Section 3.3 presents psychological theories of decision making that incorporate the time to reach the decision. Section 3.4 outlines applications to web usage. Section 3.5 provides some conclusions and discussion about future research.

## 3.2 Theories of Preferential Decision Making

Preferential choice [72] in psychology relates to the inherently subjective preferences that arise as a mixture of thoughts, motives, emotions, beliefs and desires. Subjectivity comes from the fact that the preferences of a particular individual are not easily evaluated against objective criteria without knowledge of the individual's goals. In economics, preferential choices are defined in term of rationality's consistency principle (e.g. "utility maximization"). Nevertheless, people are imperfect information processors and limited in knowledge, computational power and time availability. Furthermore, violations of rationality principles have been observed in many experiments [72]. Further extension and replacement via traditional axiomatic economics has been proposed under the name of "NeuroEconomics" [31], which incorporates psychologically-based preferential choices for describing the behavior of economic agents. This section discusses classical preferential choice theory, in relation to web usage, and proposes other promising approaches to be explored.

### 3.2.1   Tversky's Elimination by Aspect (EBA)

Tversky's model, stated in 1972, describes an algorithm for defining the decision regarding a subject's discrete choice. The algorithm was originated through behavioral studies on decision making. Each possible choice is categorized by a set of discrete aspects $\Pi$. By means of fixing one of them, a strict subset of all possible choices remains. EBA repeatedly fixes one discrete aspect until a unique option is left. The simplified Monte Carlo algorithm follows:

---

**Algorithm 3.2.1:**  ELIMINATIONBYASPECT($\Gamma, \Pi, \mu$)

---

**comment:** Simulate a selection by eliminating aspects from $\Gamma$.

**while** $|\Gamma| > 1$

**do** $\begin{cases} P_\alpha \leftarrow u_\alpha / \sum_{\alpha' \in \Pi} u_{\alpha'}, \ \forall \alpha \in \Pi \\ \rho \leftarrow \text{RANDOM}(\{P_\alpha\}) \\ \Pi \leftarrow \Pi \backslash \{\alpha_\rho\} \\ \Gamma \leftarrow \{x \in \Gamma | x_\rho = \alpha_\rho\} \end{cases}$

$k \leftarrow \Gamma[0]$

**return** ($k$)

---

Algorithm 3.2.1 describes how to select in probability a set of discrete options $\Gamma$. Each element $x \in \Gamma$ is a set $x = \alpha$ of characteristics $\alpha \in \Pi$ uniquely identified by a label $\rho$. RANDOM function is a random number generator of the discrete probability distribution of each element of $\Pi$ given by $\{P_\alpha\}$ and considering its utility assignation $\mu_\alpha$.

This algorithm (or heuristic) is called a "non-compensatory" strategic rule, since only one aspect is selected each time. In its first formulation it does not consider a previous order of attributes, but can be generalized using conditional probability, resulting in a hierarchy of aspects [92]. "Loss of aversion" is modeled by the introduction of an asymmetric function in the domain of losses [93]. This non-compensatory model of decision making has been used for marketing purposes, as in forecasting client choices based on product characteristics [30]. This is the case of web buyers of products like digital cameras, which differ in several technical attributes, and on web sites in which comparative tables are present.

### 3.2.2   Prospect Theory

This theory appeared as a critique to expected utility theory (EUT) [43] as a descriptive model of decision making under risk. The approach considers behavioral characteristics such as "loss of aversion" from the field of psychology. People tend to underweight outcomes that are merely probable in comparison to outcomes that are obtained with certainty. Another effect considered is that people tend to discard attributes that are shared by all choices (the "isolation effect"). It has been suggested [38] that this theory could be applied to customer behavior on an e-auction. An auction is the process of finding an equilibrium price and selling a good the valuation of which is difficult to estimate (e.g. art). When the internet-driven auction or e-auction process is automated, then the customer demand is satisfied. Prospect theory helps to model customer bidding behavior in such an activity for optimizing e-business revenue.

Subjects are confronted by the selection of a kind of contract called a "prospect." Prospects are quite similar to the "lotteries" of EUT, but the formalism is presented as it was conceived. A prospect can be represented by a tuple $(x_1, p_1; \ldots; x_n, p_n)$, where $x_i$ is the outcome to be obtained with probability $p_i$ on $n$ exclusive possibility. Traditional expected utility values are computed as $U = \sum_{i=1}^{n} p_i u(x_i)$ but in this approach this value is adjusted by rescaling function $v(x)$ and $\pi(p)$. Furthermore, $V = \sum_{i=1}^{n} \pi(p_i) v(x_i)$ is interpreted like an expected utility value.

The function $\pi$ reflects the impact of the probability value on the subject's appreciation of the whole prospect and must accomplish $\pi(0) = 0, \pi(1) = 1$. It is important to indicate that probabilities are never exactly perceived by the subject so other factors can influence such a function $\pi$. This function collects some behavioral observations: overestimation of rare events, overweighting and sub-certainty for the additional value of probability. Those properties have been shown to imply that $Log(\pi)$ is a monotone convex function of $Log(p)$ and with limits on values of $p$ near 0 and 1 (relaxing condition $\pi(0) = 0, \pi(1) = 1$).

The function $v(x)$ (*value function*) assigns a subjective value to the outcomes, which measure preferences with respect to a reference point reflecting gains and losses. The human perceptual machinery has been observed to respond to changes instead of absolute values. Many sensory and perceptual studies suggest that the psychological response is a concave function in the range of gains, convex for losses, and even steeper for gains than for losses. The function will have an S-shape with $v(0) = 0$ (Figure 3.1).

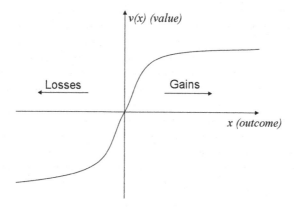

**Fig. 3.1** Value function shape

### 3.2.3 Expected Utility Theory (EUT)

This approach is related to decisions under uncertainty and risk [57] that occur for example in insurance strategy. Risky alternatives have a discrete set of
outcomes, which are described by the lottery concept. A lottery is a tuple $L =
(p_1, \ldots, p_n)$ where $\sum_{i=0}^{n} p_i = 1, p_i \in [0, 1]$ is the multinomial distribution of
outcomes occurrence.

The concept of utility in this stochastic framework is recovered as the Von
Newman-Morgenstern [96] expected utility function $U(L) = \sum_{i=1}^{n} p_i u_i$ where $u_i$
is identified as the utility value for the case i. Furthermore, one lottery is preferred
versus another by a subject if expected utility value "i" is higher. If the preference
relation between lotteries is constrained by the continuity and independence axiom,
then it can be represented as an expected utility. The continuity axiom establishes
that small changes in probabilities do not affect the preference ordering. The independence axiom establishes that convex combination with a third lottery does not
affect the preference ordering. This is called the expected utility theorem.The flexibility given by the last theorem allows the incorporation of risk aversion in the
theory as special utility functions.

Furthermore, application of this theory to the web realm has recently been performed. Expected utility was used as part of a Bayesian model [26] for analyzing browsing decisions. Navigational patterns were explored using expected utility
[44, 50] in a stochastic grammar model [12] and implemented as a recommender
system. Recommender systems based on expected utility have been applied to mobile devices [5]. Other studies [89] are more related to economics.

### 3.2.4 The Random Utility Model

A popular model in economics is the random utility model [2], where subjects are
utility maximizers but have uncertainty about the utility value. The randomness is
attributed to imperfections in the rational process. Nevertheless, subjects decide to
maximize their random utility resulting in a probability distribution for each choice.
The random component error of the utility is supposed to be additive to a classical
utility function, and dependent on the assumption of its distribution. In this way
there are at least three models assuming linear, normal (Probit) and Gumbel (Logit)
distribution.

In particular, the logit model has been proposed in [66] as a probabilistic model
for link selection in a web user's browsing behavior. Web users are proposed to be
"information foragers" where their utility is based on a TF-IDF text measure [56].
Using such a specification the SNIF-ACT browsing model has been proposed [67].

The random utility model is based on the following assumptions:

- Individuals are faced with selecting an option from a set of finite and uncorrelated alternatives. Alternative $j$ has associated attributes labeled by $k$ and quantified by the variables $X_{jkq}$ for individual $q$.
- Individuals have common motivations for making decisions.
- Individuals behave rationally (i.e. utility maximizer) and possess perfect information over possible choices.
- Each alternative $j$ is associated with a utility function $U_{jq}$ related to the individual $q$.
- The utility has a stochastic error component $U_{jq}(X) = V_{jq}(X) + \epsilon_{jq}$. The term $V_{jq}(X)$ corresponds to the traditional notion of utility depending on the feature values $\{X_{jkq}\}$ of the choice $j$. The term $\epsilon_{jq}$ is a stochastic iid term, with null average $E(\epsilon_{jq}) = 0$.
- A probability value $P_{jq}$ of choosing the option $j$ is given if the individual $q$ utility is greater than all other values $U_{jq} \geq Uiq, \forall i$. Then the probability is given by $P_{jq} = P(\epsilon_{iq} - \epsilon_{jq} \leq V_{jq} - V_{iq}, \forall i)$.
- If the density of the error term is $g(\epsilon)$ then the probability $P_j$ of choosing the option $j$ is $P_j = \int_{-\infty}^{+\infty} g(\epsilon_j)[\prod_{i \neq j} \int_{-\infty}^{V_j - V_i + \epsilon_j} g(\epsilon_i)d\epsilon_i]d\epsilon_j$

Several efforts have been made for arriving at a simpler expression for $P_j$ in order to have a simpler model. The choice probability will depend on the $g(\epsilon)$ distribution. The most-used case is the multinomial logit model (MNL) where the errors follow a Gumbel's distribution $\epsilon \sim Gumbel(\mu, \theta)$ [25], which results in the Logit distribution (3.1).

$$P_j = \frac{e^{\beta V_j}}{\sum_{i=1}^{n} e^{\beta V_i}} \tag{3.1}$$

The Gumbel distribution is an extremum class [25] that is invariant to the maximum operation, in the sense that the maximum of Gumbel's distributed variable is also Gumbel distributed. The MNL is also equivalent to the problem of maximum entropy having been restricted to an expected utility value. One important property recovered from this model is the independence of irrelevant alternatives, according to which the relative probability of selecting two alternatives does not change if we add a third independent alternative. However, this property is violated in much behavioral experimentation.

Variations of the MNL are the hierarchical logit (HL), mixed logit (ML), heterocedastic extreme value model (HEVM), and probit model (PM). All of them result from relaxation of the assumptions of the MNL.

Hierarchical logit (HL) [21] considers relaxing the independence of error $\epsilon_i$ assumption, so that independence is then established by groups called "nests". A nest is represented by the probability of choosing the whole group, and probabilities within the nest are conditionals. For each nest $S_j$ a utility-nest value is given by $V_j = \phi_j Log(\sum_{i \in S_j} e^{\frac{1}{\phi_j} V_i})$. Hence the last formula is consistent with a single-choice nest, since it collapses to the utility of the corresponding choice. Furthermore, alternatives are grouped by nested levels of independence, where probabilities are given by Bayes conditional probability. The probability of an independent choice $j$ in a nest $i$ is given by top-level nest $W_j$ which probability $P_{ij} = P_i P_{i|j} = (e^{V_i} / \sum_{l \in W_j} e^{V_l})((e^{V_j/\phi_i} / \sum_{k \in S_i} e^{V_k/\phi_i})$.

The mixed logit (ML) [90] model goes further, including in the utility description a random component that depends on the data. For instance the first proposal consists of $U_{jq} = V_{jq} + \epsilon_{jq} + \sum_k \eta_{jkq} X_{jkq}$, where $\eta_{jkq}$ is a random variable with $E(\eta) = 0$ and a different pattern of correlation and heteroskedasticity.

Another extension of the MNL is the probit model [13] where "$\epsilon$" is considered distributed by normal distribution with $E(\epsilon) = 0$ and an arbitrary covariance matrix $\Sigma$. Since the difference between two normal distributions is also a normal distribution, the probability $P_i$ of choosing the option $i$ is given by $P_i = \int_{-\infty}^{V_i - V_1} \int_{-\infty}^{V_i - V_2} \ldots \int_{-\infty}^{V_i - V_j} N(0, \Sigma_{\epsilon_j - \epsilon_i}) d\epsilon$.

Finally, the heterocedastic model (HEVM) [101] can be stated on the basis of different variances for the vector $\epsilon$.

### 3.2.5  Multiple Resource Theory (MRT)

The multiple resource theory (MRT) has its origin [98] in the study of the interference of different simultaneous tasks with the subject's attention. The theory has the ability to predict differences in performance during the operation of multiple tasks. The theory considers three aspects for information processing as explained in [99]. First, the psychological process is an aspect, which has three stages (perception, cognition, and responding). Second, the perceptual resources used by the activity corresponding to the sensory mechanism (visual, auditory, tactile, olfactory) is another dimension. Finally, the kind of reasoning (subconscious, symbolic, and linguistic) is another classification. A cognitive resource is defined by specifying all three aspects. Furthermore, a particular task could use several cognitive resources. In this sense, the performance of a task decreases as the level of the usage of cognitive resources increases.

In [62] the theory was formalized in a mathematical framework based on a utilitarian scheme of mental resources usage for dealing with a multitasking problem. The level of perceptual attention to different tasks involves a utility dependent on the performance of each task that is required to be maximized. Additionally, the mental system has technical restrictions that generate a manifold called "performance operating characteristic" (POC) that reveals all possible combinations of admissible performance tasks. In this scheme the level of attention given by a subject to each task corresponds to the tangent point of the indifference utility manifold and the POC.

In [41] MRT theory was used to optimize multiple limited resources in mobile devices for best service coordination. The approach consists of balancing human-computer interaction and conflicts with mobile processing requirements. In particular, the degree of time-continuity and human-computer interaction is balanced within an optimizing algorithm for coordination.

## 3.2.6  Scent-Based Navigation and Information Foraging in the ACT Architecture (SNIF-ACT) Model

This model [67] is based on the concept of the web user as an information forager [65]. The user is considered to be an evaluator of utility (information scent) for each possible browsing action. The goal is to describe a full-fledged simulator for a web user in order to reproduce observed interaction data. It is based on the Adaptive Control of Thought-Rational (ACT-R) psychological framework [6].

ACT-R architecture [6] is a computational simulation framework that enables cognitive features. It is based on knowledge components, of which there are two types, procedural and declarative. Chunks are the declarative minimal memory container, and productions correspond to rules that fire actions when conditions are met. It does not differ much from computer language grammar [85]; actions can alter chunks and the content of production, and can execute motor action. However, for executing the set of rules (goals) a set of buffers is necessary. Such buffering areas in the brain (see Figure 3.3) have been identified experimentally. Goal storage is identified with the dorsolateral prefrontal cortex (DLPFC) that keeps track of problem solving in the main memory [63]. The chunks buffer is identified with the ventrolateral prefrontal cortex (VLPFC) [15]. Basal ganglia areas [3] are supposed to execute pattern matching for identifying production rules and executing them.

The SNIF-ACT model [67] defines production rules for navigation and knowledge chunks within the ACT-R cognitive framework. Table 3.1 lists the production rules used by the SNIF-ACT model.

**Table 3.1** Some of the simplified production rules for SNIF-ACT navigational behavior

| | Rule | Condition | Action |
|---|---|---|---|
| 1 | Use-Search-Engine | Goal is Start-next-patch And Search Engine name is in memory | Push goal Use-Search-Engine with the given name |
| 2 | Type-Url-to-Go-Site | Goal is Start-Next-Patch And a URL is in memory | Push goal Goto-Site with the URL |
| 3 | Goto-Search-Engine | Goal is Use-Search-Engine And Search Engine name is in memory | push Goto-Site with the given name |
| 4 | Search-Using-Search-Engine | Goal is Use-Search-Engine And search terms are in memory | Enter terms in search engine pop current goal. |
| 5 | Start-Process-Page | Goal is Start-Next-Patch | push Process-Link |
| 6 | Attent-to-link | Goal is Process-Link | attent next unattended link |
| 7 | Read-Evaluate-Link | Goal is Process-Link | Read and evaluate the link |
| 8 | Click-Link | Goal is Process-Link | Click on the link |
| 9 | Leave-Site | Goal is Process-Link | pop the goal stack |

The idea behind the executing rules is that there are single rules that are going to be executed at any given time. However, cognition is a parallel process of simultaneous perception, where simultaneous true rules are solved by means of a utility scheme. The rule with the highest expected utility is picked. In the case of a web user the utility is given by the information foraging theory [66], by the notion of information scent. This model uses random utility theory for probability calculation. Furthermore, other models use as a utility the TF-IDF [56] measure over textual content [76].

### 3.2.7  CoLiDeS Model

The Comprehension-based Action planning and Instruction taking [49] (CoLiDeS) model is a simulation framework for web user browsing based on the information of the visited pages. CoLiDes follows labels according to rules, and incorporates an attentional value to different actions on the web site. It is roughly a two-nested cycle of link-clicking and link-selection actions. The stages can be separated into a goal-formation phase, attention phase, and action-selection phase. Goal formation is the initial stage where the purposes of the browsing activities are set, hence information goals are defined. In the attention phase, the web page is parsed into sub-regions, and then evaluation through semantic analysis is performed and the next phase is targeted to an interesting region. In the action-selection phase elements of the region are measured and selected according to goals. If satisfactory constraints are reached then the next action is selected.

In [48] it was stated that CoLiDeS and SNIF-ACT are complementary models, since both use an orthogonal cognitive approach. CoLiDeS considers information exploration and attention focusing on content, while SNIF-ACT explains successfully the variance in a small set of pages by strong rules.

### 3.2.8  Other Models

The Method for Evaluating Site Architecture (MESA) [61] simulates navigation by label searching within web pages. Each web page has lists of links that have an associated label, and there is a perceived likelihood for each label. Using this likelihood link, choosing could be simulated by a simple Monte Carlo technique. The whole browsing activity on the web site is then simulated page by page.

Previous models discard much of the other non-link information in a web page, an assumption that needs to be tested. In [45] the validity of this assumption was empirically researched. The surprising result was that empirically, navigation on web sites that only contain hyperlink information compared to others that maintain the same hyperlink structure but more information was the same.

However, the process of scanning a web page is an important stage in the browsing process. In [22] a simulation framework for eye movement during information scanning on a web page was proposed. The model considers both semantic, visual information, and memory models for predicting attention focusing. The model uses latent semantic techniques (LSA) for measuring the semantic similarities of each segment of the page (block). Artificial eyes are modeled by position on XY coordinates on a hierarchic block level from the whole page to a single word. The artificial eyes are simulated by scanning the page several times following a given curve. The memory process consists of accumulating measurements of the scanning time by block on an integration map. Other studies [24] raise questions about the visual complexity of a web page.

The aim of these studies is to try to understand human behavior for commercial purposes. However, the assumption that navigational support has an influence on web user behavior must be tested. Fortunately, in [29] they attempt to empirically test some navigational insights to really improve the web user experience.

## 3.3 Neurocomputing Theories

The brain, particularly the human one, is one of the most complex systems that science has attempted to describe. Having more that $10^{11}$ neurons and $10^{14}$ connections, how does this machine achieve consciousness? Neuroscience takes this question into account by attempting an answer based on neuronal mechanisms in the brain. This is a much more fundamental point of view compared with the previous symbolic-based theories. In this case a much higher complexity is exhibited since consciousness is a time-dependent phenomenon, and macroscopic descriptions depend on an enormous number of microscopic events.

Web usage can be explained by modern decision-making theories, and connectionist models are well suited to describe the time used for decision and the sequences of pages that a web user follows. A high degree of regularity has been observed on web user page sequence distributions [40], so a rule seems to govern browsing behavior. We present psychology-based theories of decision making in an attempt to depart from first principles for explaining such an observed behavior.

Current connectionism is synonymous with "computational neuroscience." This has as a definition "to study the brain as a computer, and using the computer to study the brain." The hypothesis is that neuronal electric potential variations relate with states that codify sensory feelings and abstraction. Over the last half century, several models have been proposed for explaining the reaction time of a decision. Perceptual choice is a research area where the processing and context of alternatives is studied. It considers as basic processing concepts the accumulation (i.e. integration) of perceptual data values and controlling the criteria for such values.

This section relates dynamic theories of decision making with psychological insight. In this sense the time dimension plays an important role as a variable for the decision process. We present the state of the art of neurocomputing decision theory and end with an application.

### 3.3.1  Neuroscience Foundations

The first model of the neuron dates from 1930 [39], where electrical potential in nerves was described as an accumulation of excitatory inputs from other neurons, and which theory required that a given threshold slope was needed for obtaining a response. Learning principles based on neural activity were presented in 1949 by Hebb [36] establishing that repeated excitation of a neuron by another strengthened over time augmenting the efficiency of the process. This process inspired the model of artificial neural networks in artificial intelligence. The strength of a connection between neurons is represented by a number called "weight." This value represents the fraction of activity exchanged from one neuron to another. Then Hebb's learning rule stated that such weights change for learning purposes.

Decision-making reaction time as a discrete stochastic process was covered by Stone in 1960 [86, 52]. An empirical discrete model was proposed for describing a choice's reaction time together with agreement on the psychological basis and experimental data. The proposed algorithm was based on the famous Wald's sequential probability ratio test (SPRT) for deciding between two alternatives. It is a sequential sampling model where noisy information is accumulated over time (see Figure 3.2), in which irrelevant fluctuations are expected be averaged to zero.

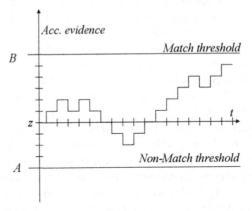

**Fig. 3.2**  An optimal Wald test sampling as a discrete random walk

The subject perceived noise evidence ($x$) about two independent choices. Two thresholds ($A, B$) corresponding to each choice, are given and used for deciding the option. The value $S$ correspond the accumulated log ratio between probabilities to decide between (0) and (1), whixh results in $S = \log(p_i^0(x)) - \log(p_i^1(x))$. The test corresponds to identify if $S$ reach the threshold $A$ for accepting 1 or if it hit $B$ then 2 is accepted, otherwise a new accumulation step should be performed. This procedure was proved to be optimal in the sense that the average number of steps (time to decide) needed for making the decision was minimal with respect to any other mechanisms [97]. Furthermore, this simple mechanism 3.3.1 requires the knowledge of the probability $p_i(x)$. Threshold $(A, B)$ has an interpretation as either an error of type I or a false positive rate $(\alpha)$ and type II or false negative rate $(\beta)$. In this case such a threshold can be estimated by $A = \beta/(1 - \alpha)$ and $B = (1 - \beta)/\alpha$. The algorithm 3.3.1 returns the choice $k \in \{0, 1\}$ and the time used for deciding (or the number of steps) $i$. Despite the statistical assumptions of the nature of this model, it was the basis for the understanding of further stochastic theories.

---

**Algorithm 3.3.1:** SEQUENTIALPROBABILITYRATIOTEST($p^0(x), p^1(x), \alpha, \beta, \{x_i\}$)

---

**comment:** Execute SPRT process for the sequence of observation $\{x_i\}$.

$S \leftarrow 0$
$i \leftarrow 1$
$B = (1 - \beta)/\alpha$
$A = \beta/(1 - \alpha)$
**while** $(S > A)$ **or** $(S < B)$
$\quad$ **do** $\begin{cases} S \leftarrow S + \log(p^0(x_i)) - \log(p^1(x_i)) \\ i \leftarrow i + 1 \end{cases}$
**if** $(S \leq A)$
$\quad$ **then** $k \leftarrow 0$
**if** $(S \geq B)$
$\quad$ **then** $k \leftarrow 1$
**return** $(k, i)$

---

A rich phenomenological area of research has been developed based on an accumulative stochastic process [86, 51, 52, 7, 69]. Such psychological models of decision making and time reaction were nicely confirmed by experimental evidence of the accumulation of information in a certain area of the brain, and will be discussed in the next section.

## 3.3.2  Biological Insight on Decision Making

Neurocomputing models are inspired by what really occurs in a brain. The aim is to obtain theories that are based on physically measurable values of psychological phenomena, and thus the first class objects in the field are neurons. Neurocomputing establishes mathematical models based on neuronal experimental facts from experimental psychology. Neurons are distinguished from other cells by the experimental fact that they can be excited electrically by other neurons. Previous phenomenological models of the time to reach a decision were not based on experimental measurements of current in the brain. However, recent advances inspired the extension of sequential sampling theories to continuous accumulative processes.

Using a new method for measuring neuronal activity, a 2002 experiment on rhesus monkeys revealed how decisions based on visual stimuli correlate with the

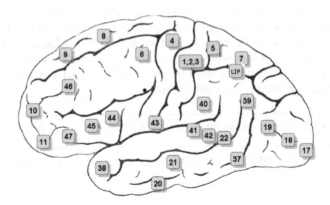

**Fig. 3.3** Human Brain Anatomy (Brodmann Areas) (modified wikimedia commons repository). Frontal Eye Field (FEF): 8, Middle Temporal Area (MT): 21, Lateral IntraParietal area (LIP), DorsoLateral PreFrontal Cortex (DLPFC):9, DorsoLateral PreFrontal Cortex (VLPFC): nearly 47, Basal Ganglia: behind 43-41, Optic tectum or Superior Colliculus (SC): behind 22 inside the Thalamus, Primary Somatosensory Cortex: 1-2-3, Fusiform Face Area (FFA): near 37, ParaHippocampal Area (PPA): Inferior Frontal Gyrus (IFG): 11.

middle temporal area (MT) of the brain [73]. The experiment consisted of presenting a screen with random moving points, where subjects had to decide to move their eyes in one of two directions. Each monkey was implanted with electrodes recording the activity of 54 neurons in the lateral intra-parietal cortex (LIP) and in the MT. Figure 3.3 presents a rough description of the anatomy of a human brain, where the LIP region on the brain's right upper corner and the MT area in the bottom zone are present in most mammals. Eye movements were tracked on a horizontal and vertical plane. The subjects learned to respond to a reward of fruit juice if they indicated by eye movement the correct direction of movement of points on the screen. The screen was presented with a random movement of dots and the reaction time of the decision was recorded. The experimental data was interpreted at a physiological level. Before reaching a decision the increase and reduction of the spike rate was observed. When the activity had reached a threshold value and the decision process was complete, the monkey responded 50 msec later, and the LIP activity terminated. The process suggested an accumulation of information toward a threshold. The LIP area is then a temporary storage area of partially-recorded information, and is in charge of processing the decision by means of a kind of sequential analysis processing analogous to Wald's SPRT (See section 3.3.1).

The MT area appears to be a temporary storage and processing device for visual information that is transmitted to the LIP area for further processing. Earlier studies [14] showed a linear correlation of visual stimuli on the neural activity of the visual cortex MT. The experiment was the same as [73], measuring neuronal activities by mean of electrodes.

More recently in [35], the same experiment was modified to induce neuronal stimulation of the monkey brain's LIP area. The purpose of this external stimulation was the observation of an artificially-induced decision on the subject. The experiment was set up first identifying those neurons presenting activity and which correlated with each decision choice. Neurons associated with one alternative were micro-stimulated by electrodes. Half of the measurements were taken with the bias of micro-stimulation of the correct choice and the other half of the wrong one. The result was that micro-stimulation produced an effective bias on the monkey's choice, but if the evidence of the moving dots was strong enough (i.e. a clearer movement) then the bias was smaller. Furthermore, the micro-stimulation effect was small, but significant. Another effect noticed was an increased time taken for the wrong decision and a faster response for stimulation with the correct answer. MT area (Figure 3.3) micro-stimulation was also explored but the results were weaker than for LIP. Choices seemed to have been affected as if the dot motion perception were distorted to the point of reaching logical inconsistency, generating artificial evidence over visual perception, but having the duration of the artificial stimuli. Finally the experiment concluded that decisions in the experiment were strongly correlated with the neural activity levels of specific neurons in the LIP that receive sensory evidence from the MT areas, using a threshold level mechanism.

Furthermore, the neurophysiology of decision making has been explored with multiple-choice decision making [23]. The experiment was similar to the previous using a screen with random dots moving principally in four or two directions. 90 electrodes per brain on the LIP area cortex were installed in monkeys trained by reward to identify the busy pattern. It was observed that it took a longer time to decide between four alternatives than two. An important effect was the reduction of the maximum activity level in the LIP threshold compared with the two-choice option. The interpretation is related to a kind of trade-off adjustment of the threshold for gaining speed in decision versus accuracy. As the number of alternatives increases, the brain needs to accumulate more information in order to make a decision and takes more time. The time usage per decision seems to be limited by an internal mechanism based on the implementation of a deadline. Previous findings about information accumulation in the LIP were thereby confirmed [35].

Moreover this decision mechanism has a correspondence with statistical efficiency and is more automatic than if driven by a higher level of reasoning. Psychologists argue that those empirical mechanisms are followed by a reliability assumption about the decision. A recent experiment [46] presented the correlation of the LIP areas with the degree of confidence. The certainty of the decision in the experiment was measured by implementing the decision to opt out with a small reward revealing the confidence degree.

The LIP area is not the only location that has been found to work as an information-integrating area. The frontal eye field (FEF) [81] has been experimentally studied to be proven as a recording, processing, and motor structure for visual discrimination tasks [84]. The superior colliculus (SC) generates saccadic eye movement and maintains a map of eye movement. It is connected topographically with the FEF area. The pool of neurons in such an area can be understood as a vector that averages neural electrical activities that finally discharge producing a particular eye's saccade [83]. SC neural activity has been highly correlated to changes in the saccadic probability [8] and its neural activity predicts eye movement time reaction. The dorsolateral prefrontal cortex (DLPFC) [47] seems to be a neural correlate with the short-term memory for object location and is correlated with visual decision. The DLPFC area has a response similar to that observed in the LIP, and it was suggested [47] that it acts as an integrator region for sensory information. In vibrotactile frequency discrimination experiments where subjects compare frequencies of vibration, it has been explored whether the neural activity of the primary somatosensory cortex (or sensory homunculus, see Figure 3.4) involves neural information accumulating for firing decisions [55].

**Fig. 3.4** Sensory homunculus on the primary somatosensory cortex (wikimedia commons repository), it corresponds to a tactile-sensory body map. The figure presents a slice of the primary somatosensory cortex and its correspondence with a body part.

According to [32] perceptual decision making consists of three stages of processing in the brain. First, a representation of sensory evidence is carried after recollection from the sensory system. Second, integration of the available sensory information over time is performed on an appropriate buffer. Finally, there is a comparison of the evidence to a decision, or in some cases a threshold for triggering a decision. Such a biological process 3.3.2 is very similar to the algorithm 3.3.1. The algorithm returns the time ($i$) used for decision and the decision made.

---

**Algorithm 3.3.2:** SIMPLEPERCEPTUALDECISION($\{x_i\}$, *Decision_Rule(S)*, *Decide(S)*)

---

**comment:** Brain executing perceptual decision making using sensory data sampling $\{x_i\}$.

$S \leftarrow 0$

$i \leftarrow 0$

**while** !*Decision_Rule(S)*

   **do** $\begin{cases} S \leftarrow S + \psi(x_i) \\ i \leftarrow i + 1 \end{cases}$

$k = \text{Decide}(S)$

**return** $(k, i)$

---

Recently in [37] it was proposed that perceptual decision making is composed of several complementary and overlapping systems working in parallel. Such a study is based on neuroimaging techniques for a face-recognition-task experiment, and involves the fusiform face area (FFA), the parahippocampal place area (PPA) that accumulates sensory evidence over time, the DLPFC for computing decision variables, the anterior insula (AINS) and inferior frontal gyrus (IFG) for detecting perceptual difficulty or uncertainty and for firing more attention resources for processing. The decision variable [32] is defined as the computed value in the cortex (e.g. DLPFC) based on evidence, experiences that fire a decision. The whole system is monitored by the posterior medial prefrontal cortex (PMPFC), which detects errors and the need for adjustment for maximizing performance. Such a system [64] does not act in a hierarchical way, but some steps need to be processed first (e.g. sensory evidence accumulation) before it arrives at the motor region. This model differs from previous ones that claim a hierarchy in perceptual decision making, establishing first sensory evidence collection, then accumulation (or integration) of information, and finally categorization for a motor response [91].

On the basis of those previous experimental facts one important hypothesis can be stated. Brains are stochastic information accumulators, which decide under specific decision rules. Further sections describe specific models based on such a hypothesis. However, stochastic accumulation models were proposed in the past in the absence of any knowledge of modern experiments directly measuring neural activity by electrodes or neuroimaging techniques.

## 3.3.3   Wiener Diffusion Process

Earlier in 1970, Emerson [28], and later in 1978, Ratcliff [69] proposed a continuous-time decision theory for memory retrieval using a simple "random walk" process that was tested with experimental data. The model reproduced the distribution of reaction times and error latency of subjects performing the task of recognizing items. The experiment consists of a set of single words that are presented to subjects, who are tested on the list by answering "YES/NO" depending on recognition. Time to reach a decision is also recorded. The stochastic model considered for each choice was a random walk with drift (Equation 3.2), but the vector $X = [X_i]$ does not represent any biological value. It is interpreted as a value that accumulates evidence.

$$dX_i = I_i dt + \sigma dW_i \qquad (3.2)$$

The last stochastic process is called a Wiener process (e.g. [71]). It starts at $t = 0$ at $X_i = z > 0$. Each time a comparison over two thresholds $((0, a))$ is made, a decision can be considered to have been made when any comparison terminates in a hit with an upper limit $(a > z)$, or all processes reach a lower limit $(0)$, stopping without a decision. The memory retrieval model associates each coordinate $X_i$ with an item to be identified by the subject. The process ends when a coordinate $i^*$ reaches the threshold $a$, in which case the recognized item is $i^*$ and the answer is "YES" otherwise all coordinates must have reached the 0 lower limit answering "NO".

Simulation of this process can be performed by means of Monte Carlo techniques (e.g. [80]). The algorithm 3.3.3 is based on the Brownian motion simulation "$Brownian(\delta)$" in a time interval $\delta$. The condition $Reach\_a\_Decision(X, a)$ is true only if $(\exists i^* \, X_{i^*} \geq a)$ or $(\forall i \, X_i = 0)$. The decision recognition is "YES" if $(\exists i^* \, X_{i^*} \geq a)$ and "NO" if $(\forall i \, X_i = 0)$, which is the result returned by $Decide(X, a)$. The vector $I$ is obtained by simulating a Gaussian variable with mean $\mu$ and variance $s$.

---

**Algorithm 3.3.3:** RATCLIFFMEMORYRETRIEVAL$(K, z, a, s, \sigma, \delta, \mu)$

---

**comment:** Recognizing item $K$ in memory.

$X \leftarrow z$
$t \leftarrow 1$
**while** !$Reach\_a\_Decision(X, a)$
$\quad$ **do** $\begin{cases} I \leftarrow Gaussian(\mu, s) \\ \textbf{if } (\exists i^* X_{i^*} \leq 0) \\ \quad \textbf{then } X_{i^*} \leftarrow 0 \\ \quad \textbf{else } X \leftarrow X + I\delta + \sigma Brownian(\delta) \\ t \leftarrow t + 1 \end{cases}$
$k = Decide(X, a)$
**return** $(t, k)$

---

Once a decision is ready, a response triggers another mechanism for response. The entire process is driven by the $\{I_i\}$ set of drift, the alternative with the bigger value having the best possibility of being chosen. In this study [69] the drift value is called "relatedness" of the alternative to the problem decision, which integrates partial information about the supporting evidence for the choice. Relatedness is considered to be stochastic and following a Gaussian distribution. Fortunately this model has exact solutions for decision time and distributions [69] that are used for parameter-fitting purposes using the maximum likelihood method.

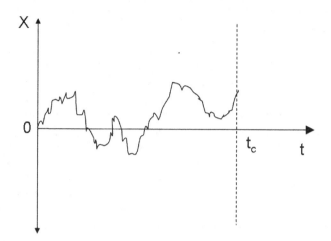

**Fig. 3.5** Weiner Process with timeout $t_c$

Another criterion for decision definition is establishing a timeout limit. Under this consideration the process loosens the variability of the time to decide, replacing it by the timeout limit parameter of the model $t_c$. Figure 3.5 describes the evolution of the variable $X$.

The model parameter magnitudes "$I_i$" reflect the difficulty in reaching a decision, since for smaller values the time to make a decision in the threshold's model can be a larger value. In the absence of noise $\sigma = 0$ and with no distribution for $I$, the model could be interpreted as the subject reaching the decision with the maximal $I_i$ value. The equation 3.2 becomes deterministic and it is solved by simple integration. In this case the time used to reach a decision "YES" is deterministic and $t_c = MaxI_i/(a-z)$.

The performance of models is measured by the error rate, which is then considered as a mechanism of decision under noisy evidence. The error rate is a function of the model parameter $I_i$, $\sigma$ and $a$. In the case of a model with a threshold and two choices ($I = I_1 - I_2$) the error rate is $P(a) = 1/(1 + e^{\frac{2Ia}{\sigma^2}})$. In a case with a timeout $P(t_c) = \int_{-\infty}^{-I\sqrt{t_c}/\sigma} \frac{1}{\sqrt{2\pi}} e^{-\frac{I^2}{2}}$ is the error rate.

This model was extended [70] to allow some parameters to be Gaussian random variables. In this case additional parameters like the mean and variance of the Gaussian distribution need to be adjusted. This change improves the degree of fit to the experimental results. Additionally, the starting point for Brownian motion $z$ is considered to be an additional stochastic parameter of the model and is considered to follow a uniform distribution.

A boundary mechanism was considered for modeling threshold stopping time. In this case an absorbing boundary was used for terminating the process (no time pressure) [60]. Reflecting boundaries were also used for controlling the integrated information, as a lower limit on accumulated value.

### 3.3.4    Decision Field Theory (DFT) and the Ornstein-Uhlenbeck Process

Busemeyer et al. in 1993 [19] published the decision field theory (DFT), a mathematical description of decision making with a cognitive basis. This framework aimed to provide an explanation of experimental violation of some common economic decision-making principles such as stochastic dominance, strong stochastic transitivity, and independence over alternatives and serial position effect on preferences. Furthermore, it explained speed-accuracy trade-off effects in decision making, the inverse relation between choice probability and decision time, changes in the direction of preferences under time pressure, a slower decision time as compared with an approach including conflict avoidance, and preference reversal between choice and selling-price measures of preferences. The first presentation of this theory considered the mentioned characteristics and constructed it on this empirical evidence.

The theory is described by a time-dependent vector $X = [X_i(t)]_{i=1}^n$, a n-dimensional preference state vector [17], which coordinates $X_i \in \mathbb{R}$ relative to the alternative $i$. At $t = 0$ such a vector represents preferences before any information about actions is considered, such as memory from previous experience ($\sum_i X_i(0) = 0$), but if a decision doesn't have an antecedent then the process starts from zero ($X_i = 0 \forall i$). The decision-making process consists of the time evolution formulated in discrete time step $h$ 3.3.

$$X(t) = S \cdot X(t - h) + V(t) \tag{3.3}$$

$$S(h) = Id - h\Gamma \tag{3.4}$$

$$X(t) = \sum_{k=0}^{T} S^k \cdot V(t - kh) + S^T \cdot X(0) \tag{3.5}$$

Where S is the $nxn$ feedback matrix, $V(t)$ is a noise input vector called the valence, and $Id$ is the $nxn$ identity matrix. When $h \to 0$, $X$ is a continuous Ornstein-Uhlenbeck process, and if $\Gamma = 0$, then it is reduced to the Weiner process. The matrix $\Gamma = [\gamma_{ij}]$ is symmetrical ($\gamma_{ij} = \gamma_{ji}$) and diagonal elements are equals ($\gamma_{ii} = \gamma$). If such a matrix represents intensity of connection, then diagonal elements represent self-feedback and off-diagonal elements represent lateral inhibitory connections that can vary over conceptual distance between choices.

The stochastic valence vector is decomposed into $V(t) = C \cdot M \cdot W(t)$. The matrix $C = [c_{ij}]$ is called the contrast matrix, which is designed to compute the advantage of an action relative to others, and its value $c_{ij} = \delta_{ij} - 1/(n - 1)(1 - \delta_{ij})$ ($\delta{ij}$ is the Kronecker delta function. The matrix $M = [m_{ij}]$ is interpreted as the affective evaluation of the consequence of performing the choice $j$ but not the choice $i$. The stochastic vector $W(t) = [w_i(t)]$ corresponds to the weight of the result of the attention process for each choice. For instance, if the decision is about a characteristic of a visual stimulus, then such a vector corresponds to the result of the processing of the visual cortex. Such a vector is supposed to fluctuate, representing changes of attention over time, and is considered as a stationary process with

$E(W) = \bar{W}h$. In this case the product $\text{Cov}(W) = \Psi h$ is an average evaluation of each action at a particular moment. Finally, $V$ will represent advantage or disadvantage over the average of the other action at any given moment.

DFT is based on the Weiner process including a dissipation term with parameter $\lambda$ (3.6). This results in $h \to 0$ as seen in [17, 16]. The new equation introduces the property of asymptotic attraction for the larger $t$ of the variable $X$. Notice that for $\lambda > 0$ this equation has an attracting fixed point $X_i^0 = I_i/\lambda$, otherwise this point is unstable. A property acquired with the introduction of this new term is the "recency effect." Furthermore, early accumulations of information from $I_i$ are erased from the new term, and in this case recent stimuli drive the time evolution of $X$.

$$dX_i = (-\lambda X_i + I_i)dt + \sigma dW_i \qquad (3.6)$$

Thresholds are incorporated in the theory in the same way as for the Weiner process.

### 3.3.5   Leaky Competing Accumulator Model (LCA)

In 2001 Usher and McClelland [94] published a diffusion model for decision making that incorporated the dissipation term of Busemeyer's DFT [16] and a lateral inhibition effect. This new term is included for taking relative accumulated evidence into account. Such work considers a theoretical unification of concepts from cognitive-perceptual processes and the underlying neurophysiology. Furthermore, previous models were based on phenomenological observations, but the LCA model uses a neurophysiology basis for describing the decision making process.

$$dX_i = [I_i - (\kappa - \alpha)X_i + \alpha f_i(X_i) - \lambda \sum_{j \neq i} f_i(X_i)]dt + \sigma_i dW_i \qquad (3.7)$$

The equation 3.7 presented in the 2001 paper considers the following parameters. $I_i$ is an input value in favor of the alternative $i$ that is accumulated from other devices such as the visual cortex, and serves as an input to the LCA process. Those values are supposed to be constrained as $I_i \geq 0$ under neurophysiology reasoning. External input values are accumulated in the variable $X_i$ in favor of the alternative $i$. The $\kappa - \lambda$ parameter takes into account the decay [58]. The $\alpha f_i(X_i)$ term is a recurrent excitatory source coming from the unit $i$ and modulated by the function $f_i() \geq 0$. Lateral inhibition between accumulator units is controlled by the $\lambda$ parameter and considers the effect equal for all units, but is modulated by the function $f_j() \geq 0$. The accumulated values are considered biological values, such as neural activity (rate of spikes), which are then restricted to being positive. Function $f_i()$ is near to a linear neural response, and an approximation could be considered as $f_i(x) = x$. This is a difference from previous models, since the aim is that $X$ be adjusted to the real biological value of neural activity. Hence DFT and some variations of the Wiener process use variables that take negative values. The model then considers that a decision begins when the first accumulator $X_i$ reaches $i^* = \text{ArgMax}_i(X_i(t^*))$ where

$X_{i^*}(t^*) = X^*$ on the threshold $X^*$. Otherwise, if the phenomenon is time constrained (i.e. with a timeout) then the actual maximal value is used as the decision.

In the linear approximation of the $f_i$ function and time-constrained decision case, there are exact solutions for the probability density of the process without considering border condition. In [94] the process was analyzed for the case of two dimensions and proved to be an Ornstein-Uhlenbeck process with a solution in terms of $x = X_1 - X_2$ following a Gaussian distribution 3.8 with a time-dependent mean 3.9 and variance 3.10.

$$x \sim N[\mu(t), \sigma(t)] \tag{3.8}$$

$$\mu(t) = \frac{I_1 - I_2}{\kappa - \lambda}(1 - e^{-(\kappa - \lambda)t}) \tag{3.9}$$

$$\sigma(t) = \frac{\sigma}{\sqrt{\kappa - \lambda}} \sqrt{1 - e^{-2(\kappa - \lambda)t}} \tag{3.10}$$

$$\epsilon(t) = \frac{2\mu(t)}{\sigma(t)} \tag{3.11}$$

This model exhibits asymptotic behavior for the average $\mu(t) \rightarrow \frac{I_1 - I_2}{\kappa - \lambda}$) and variance ($\sigma(t) \rightarrow \frac{\sigma}{\sqrt{\kappa - \lambda}}$). For measuring the accuracy of the model 3.11, it was defined using a signal theory accuracy ratio between the separation of the mean of success and failure versus the variance [100]. This value is asymptotically $\epsilon(t) \rightarrow \frac{2(I_1 - I_2)}{\sigma \sqrt{\kappa - \lambda}}$.

### 3.3.6 Other Models

A novel approach for the mathematical modeling of decision making is the quantum dynamic approach [20]. In such a model the mathematical framework of quantum mechanics is applied for explaining decision making. Nevertheless no neurophysiology principles have been used for supporting those assumptions, but the formalism is used as a richer formulation that better explains some known violations of rational theories [68]. The new approach is known as quantum decision theory (QDT), and claims that the brain is a quantum computer.

A Markov process has been used with considerable success in explaining aspects of decision making. Quantum dynamics has many similarities with Markov processes. Quantum states are related to complex probability amplitudes $\varphi \in \mathbb{C}$, the inner products between which are transition probabilities between their respective states. The evolution rule is similar to Kolmogorov's equation for probability $P_{ij}$ transition $\frac{dP_{ij}}{dt} = Q \cdot T(t)$. In the case of quantum evolution the state evolves according to 3.12, where $H$ is the Hamiltonian operator.

$$\frac{d\varphi}{dt} = -iH\varphi \tag{3.12}$$

A basic example considers the quantum-based two-choice problem [20]. States are described by a level of confidence $l \in \{0, 1, \ldots, m\}$ that indicates a measure of the likelihood of the decision where zero is indifference. The decision is binary $+/-$. Hence, the number of states is $m + 1$, since a codification of possible states is the set of integers $\{-m, -(m - 1), \ldots, m - 1, m\}$. The system is represented by linear combinations of wave functions representing those states. The Hamiltonian matrix $H = [h_{ij}]$ for this problem is defined in 3.13.

$$h_{ij} = -j\mu\delta_{ij} - \frac{\sigma^2}{\Delta^2}(\delta_{i-1,j} + \delta_{i+1,j}) \tag{3.13}$$

If the initial state of the system is given as $\varphi_0$, then in a time $t$ the result is that $\varphi(t) = e^{-iHt}\varphi_0$. If the subject has a deadline in which to decide, the process stops and a measurement of the level of confidence is performed. A quantum measurement corresponds to a protection of the wave function of the system in the subspace that represents the measurement. In this case a choice corresponds to a projection to a positive confidence value for choice (+) and negative for choice (−). Probabilities are then inner products of those projections on the wave function.

More recently [18], this approach was applied to the prisoner dilemma and compared to the Markov approach.

Grossberg [34] in 1987 proposed a model based on neural network dynamics for explaining decision making under risk. This theory is called affective balance theory (QABT). The theory states that the stimulus is not the only factor in decision making, but that cognitive context alters the interpretation and consequent behavior. This model incorporates emotional and physiological factors. The dynamics of the theory become non-linear at each step. Nevertheless, it is presented as an example of a further enhancement of the theory.

## 3.4   A Neurocomputing Application to the Web Usage Analysis

In this section we present an application [76, 75, 78, 74] of the neurophysiology-inspired LCA model to the simulation of the web user.

Two different ways exist for describing a natural phenomenon, generic methods and first principle-based methods. Generic methods relate to general mathematical models that are fitted to available data in order to predict and describe observations. Data mining techniques are based on generic methods for discovering unexpected regularities in unknown data. Machine learning algorithms are further optimized for handling large amounts of data. Furthermore, without any preconceived understanding of the observed data, data mining techniques constitute an outstanding tool for analysis. If on the other hand, first principles are known to rule the observed data, then a much better approximation of the phenomenon can be obtained if the mathematical model adopts such conditions.

First principle mathematical models do not discard the use of generic machine learning techniques. The first principle description only partially describes the real behavior of the phenomenon, since a theory of everything has yet to be discovered, or perhaps never will be. Effects not covered by the theory are mainly represented by parameters of the models. For instance, the Newtonian law that describes a harmonic oscillator requires the value of a mass and spring constant. Such parameters represent atomic interactions that are abstracted and simplified into one adjustable value. Machine learning techniques help to adjust such parameters using available data. Accordingly, both approaches are intrinsically complementary for better modeling of natural phenomena.

As presented, the human decision process is a natural phenomenon which has been studied by the field of psychology for decades. First principle mathematical models have been tested experimentally and related with the physical description at the level of neural electrical activity in the brain. Human decision-making behavior is described by a stochastic process that has many parameters. This theory is applied to each navigational decision process of a web user confronted by a web page, who is considering each hyperlink as a possible choice. The proposed model predicts a probability distribution for each choice, as well as the time taken to reach a decision.

### 3.4.1   The Time Course of the Web User

The current models of a web user's navigational behavior have related more to describing the visited combination of pages than the timed sequence of them. We interpret the dynamic model as describing the forces involved in link selection or otherwise leaving the site, by recovering the jump sequence between pages.

A web user's visit starts when he/she reaches a page from a chosen web site, which is defined as the initial page on the trail of page sequence (also called a "session.") Initial page arrivals have a very typical distribution (Figure 3.6), and the implicit mechanism for reaching such pages could be from a search result (e.g. Google), navigator bookmarks, navigator suggestions and links from other web sites. As an example over a period of 15 months the first page distribution on the http://www.dii.uchile.cl web site had the distribution shown in figure 3.6. The root page has nearly 35% of all first visits, while the rest have scarcely less than 5%.

Intuition indicates that the first page will define the semantic of further navigation of a user on a web site. However, other factors appear to influence this decision. In the case of a university computer center, the first browser page is usually set to a given department's page, in which case the default browser's page visits do not influence the semantic of further visits. Many first-time visitors to the web site (beginners) will typically try a first page that probably does not clearly represent their purposes, since they are unfamiliar with the site. We can conclude that most of the web user's intentions are contained in the "jump" decisions between pages, rather than the first page visited. Furthermore, visits consisting of a single page could be discarded from the analysis, and first page distribution is exogenous to this model.

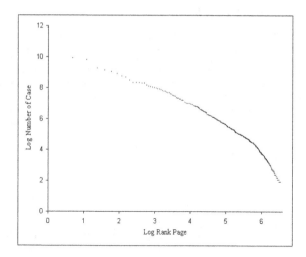

**Fig. 3.6** Log-distribution of the ranked first page of a session (source: web site http://www.dii.uchile.cl)

During a visit to a web page, the web user has to decide which link will be selected, or finally leave the web session. Web users are modeled as information foragers [66]. Furthermore, they experience a degree of satisfaction with consuming the information included on web pages. This idea is influenced by the economic theory of utility maximization, where a web user is a consumer of information, selecting the link that most satisfies him. However, a model which only considers this dynamic factor would produce a web user that never stops navigating. For example the random surfer [10] model is a naive description of a web user that has no interest at all in the web page content. Furthermore, a random surfer does not rely on web page content, he/she only uniformly decides on the next link to follow, or leaves the site with probability $d$. Moreover, this probability $d$ seems to be a constant over all sites. If we include the exit choice, a new option corresponding to leaving the web site is incorporated. As a first approximation this option should be tested on real data.

The previous description will not be complete if decision time is not considered in the evolution of the system. The perceptual choice theory describes the time to decide based on a stopping time for the diffusion model. That conforms to the specifications for the navigation of an artificial agent on the web, where at each page it decides which option (either click on link or leave) to choose according to the LCA model. Nevertheless LCA depends on the perception of the text on the web page, which would correspond to a higher cognitive area's probabilistic processing results. Such a kind of probability value has already been explored in economics according to the random utility model, where in this case a utility for content will be explored.

### 3.4.2   A Model From Psychology

The LCA model is used for describing the time evolution of probabilities regarding a web user's decision making. This model has the advantage of being easily extended by plugging stochastic force into the evolution equation in order to capture new behavior.

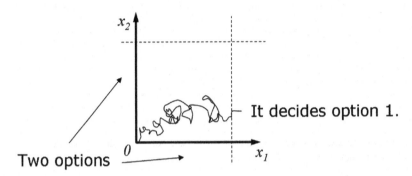

**Fig. 3.7** A diffusion-based decision making model. The first coordinate to reach the threshold corresponds to the final decision.

Models like the LCA stochastic process have a long history of research and experimental validations, most of which have been carried out in the last 40 years [52, 82, 86, 69]. However, few engineering applications have been proposed up to now. This work assumes that those proven theories on human behavior can be applied and adapted to describe web user behavior, producing a more effectively structured and specific machine learning model. The approach consists in applying the LCA model to predicting the web user's selection of pages (session). This proposition was based on experimental validation. A web user faces a set of discrete decisions that corresponds to the selection of a hyperlink (or leaving the site).

The LCA model is applied to simulate the artificial web user's session by estimating the user's page sequences, and furthermore by determining the time taken in selecting an action, such as leaving the site or proceeding to another web page. Experiments performed using artificial agents that behaved in this way highlighted the similarities between artificial results and a real web user mode of behavior. Furthermore, the performance of the artificial agents was reported to have statistical behavior similar to humans. If the web site semantic does not change, the set of visitors remains the same. This principle enables the predicting of changes in the pattern of access to the web page, which in turn are related to small changes in the web site that preserve the semantic. Web user behavior could be predicted by simulation, and then services could be optimized. Other studies on ant colony models [1] relate directly to general-purpose clustering techniques.

The neurophysiology of decision making [94, 11] and the random utility model of discrete choices [59] are considered to model the web user's behavior. In the field of mathematical psychology, the leaky competing accumulator (LCA) model describes the neurophysiology of decision making in the brain [94]. It corresponds to the time description of the subject neural activity of specific zones $\{i\}$ in the brain.

$$dX_i = (F_i^D + F_i^C + F_i^E)dt + \sigma dW_i \tag{3.14}$$

$$F_i^D = -\kappa X_i \tag{3.15}$$

$$F_i^C = -\lambda \sum_{j \neq i} f(X_j) \tag{3.16}$$

$$F_i^E = I_i \tag{3.17}$$

For each decision $i$ a region in the brain is associated, which has a neuronal activity level (NAL) activity $X_i \in [0, 1]$. If a region $i_0$ reaches an NAL value equal to one, then the subject makes the decision $i_0$. The NAL's $X_i$ are time-dependent, which dynamic is stochastic as shown in the equation 3.14. Several forces $(F^D, F^C, F^E)$ drive the system including the stochastic force $\sigma dW_i$.

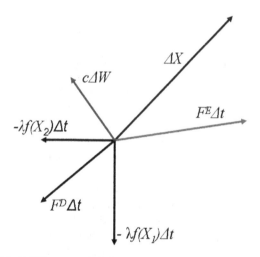

**Fig. 3.8** Forces of the LCA's system in 2-D

The forces are interpreted as: $F^D$ is the dissipative force in the case and is responsible for vanishing memory if no other interactions are present, $F^C$ is the competing term related to inter-inhibition of neural connection observed in real network tissue, and $F^E$ corresponds to the likelihood value that other cognitive processing gives to each choice.

$F^E$ is called the evidence term. An experiment on visual discrimination with random moving points explores the way in which the visual cortex furnishes the likelihood value for the direction of the movement. $F^E$ is the representation of such input

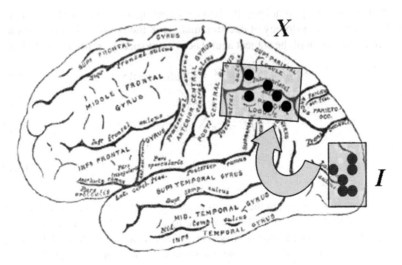

**Fig. 3.9** Evidence Forces from the Visual Cortex (modified wikimedia commons repository)

from sensory devices. Those brain's pre-processing likelihoods directly reinforce the neural activity $X$.

The parameters of the theory are interpreted as: $\kappa$ is a dissipative coefficient, $\lambda$ is related to competitive inhibition between choices, $I_i$ is the supporting evidence of choice $i$ and $\sigma$ is the variance of the white noise term $dW_i$. The function $f(.)$ corresponds to the inhibition signal response from other neurons, usually modeled as a sigmoid (near to linear) or in this case linear ($f(x) = x$). The parameter $I_i$ in the LCA theory is interpreted as likelihood values regarding the importance of choice $i$ for the subject. Other diffusion models have been proposed, but all have been proven to be equivalent to LCA [11].

Web users are considered stochastic agents [76, 75, 78, 79, 77]. Those agents follow LCA stochastic model dynamics (Equation 3.14), and maintain an internal state $X_i$ (NAL's values) with some white noise $dW_i$. The available choices, including the probability of leaving the web site, lie in the links on a web page. Agents make decisions according to their internal preferences using a utilitarian scheme as shown in the next section 3.4.3.

An important observation about the equation 3.14 is that it resembles Newton's equations. As a matter of fact it is called a Langevin's Equation, since it incorporates a stochastic force $\sigma dW$. The variable $X$ is measured as the rate of electric signal spike per second, since it could be considered as a kind of velocity. Furthermore $\frac{dX}{dt}$ can be understood as acceleration. Forces in this model can easily be included as an additive term $Fdt$, opening increased possibilities of expanding the decision-making model's influence on other cognitive phenomena.

### 3.4.3   A Random Utility Model for Text Preference

The vector $(I)$ is the main force that drives the decision system (3.14). Furthermore, we model those values as proportional to the probability $P(i)$ of the discrete choices ($I_i = \beta P(i)$), which are usually modeled using the random utility model. Discrete choice preferences have been studied in economics to describe the amount of demand for discrete goods where consumers are considered rational as utility maximizers.

$$P(i) = \frac{e^{V_i}}{\sum_{j \in C} e^{V_j}} \tag{3.18}$$

The logit model has been successfully applied to modeling a user's search for information on a hypertext system [66], resulting in improved adaptive systems. The utility function should depend on the text present in links that the user then interprets and by means of which he/she makes the decision.

Intuition about the functional form of a similarity measure tells us that the similarity between vectors $TF/IDF$ should be invariant under expansion ($sim(x, y) = sim(\lambda x), \lambda y)$) because of the lack of units of the component. The simplest functional form to have this property of homogeneity of degree 0 is the cosine $sim(x, y) = cos(x, y) = (x \bullet y)/(\|x\|\|y\|)$ where "$\bullet$" is the dot vector product.

Hence the assumption is that each agent's link preferences are defined by its $TF/IDF$ text vector $\mu$ [56]. The $TF/IDF$ weight $\mu_k$ component is interpreted as the importance for the web user of the word $k$. Furthermore, an agent prefers to follow similar links to its vector $\mu$. The utility values (equation 3.19) are given by the dot product between the normalized $TF/IDF$ vector $\mu$ and $L_i$ that represents the $TF/IDF$ weight text vector associated with the link $i$.

$$V_i(\mu) = \frac{\mu \bullet L_i}{|\mu||L_i|} \tag{3.19}$$

The resulting stochastic model (equation 3.14) is dependent on the parameters $\{\kappa, \lambda, \sigma, \beta, \mu\}$ and the set of vectors $\{L_i\}$. The first four parameters must be considered as universal constants of neurophysiology, yet the $\mu$ vector is an intrinsic characteristic of each web user. Thus, the real web user's mode of behavior as observed on a web site corresponds to a distribution of users.

A web user is considered a stochastic agent that "surfs" the Web according to the stochastic decision rule 3.14 with preferences defined by 3.19. Parameters like $\{\kappa, \lambda, \sigma, \beta\}$ represent the physiological constants of neural tissue that every human must share. However the vector $\mu$ should be interpreted as the web user text preference at the moment of visiting the web site. We are assuming the web user does not change his/her intention during a session and leave the web site according to constant probability. The vector $\mu$ drives each web user's behavior. In this model web user profiling is in direct relation with the $\mu$ vector distribution.

### 3.4.4  Differential Equation for Probability Distributions

Numerical methods for calibrating the proposed model need to be tailored to fit the current mathematical description. Well-known statistical models are based on definite probability distributions (normal, exponential, etc.) which are dependent on the parameters. Maximum-likelihood methods maximize the probability of finding the parameters of such probability distributions in observed data. This technique is called "parametric inference". However the present model is far from having a simple or exact solution, despite its easy simulation. The problem becomes how to find the probability distribution function and the parameters that describe the model. This kind of problem falls in the category of "non-parametric inference". For those purposes the mathematical behavior of probability distributions must be analyzed in detail.

The statistical description of the system is analyzed based on The Kolmogorov forward (or Fokker Plank) differential equation on the following set (equations from 3.20 to 3.24).

$$\frac{\partial \phi}{\partial t} = \sum_i \frac{\partial}{\partial X_i}[-\phi F_i + \sigma^2/2\frac{\partial \phi}{\partial X_i}] \tag{3.20}$$

$$F_i = -\kappa X_i - \lambda \sum_{j \neq i} f(X_j) + I_i \tag{3.21}$$

$$\phi(X,t) = 0, \ \forall t > 0, \ X \in \Psi \tag{3.22}$$

$$\hat{n}(X) \bullet (\sigma^2/2\nabla\phi + \phi F) = 0, \ \forall X \in \Delta, \hat{n}(Y) \perp \Delta, t > 0 \tag{3.23}$$

$$\phi(X,0) = \delta(X) \tag{3.24}$$

The function $\phi(X,t)$ is interpreted as a joint probability density at time $t < T$, where $T$ is the time of the first hit to the boundary $\Psi = \bigcup_i \Psi_i$, $\Psi_i = \{X|X_i = 1\}$. The restriction 3.22 indicates considering a case in which the variable $X$ has never reached the $\Psi$ barrier. Furthermore, $\phi(X,t)$ is the probability density of the subject to have a neural activity level of $X$ without already having reached a decision. This absorbing barrier $\Psi$ is compounded by perpendicular planes $\Psi_i$ that have the coordinate $X_i = 1$, which represents the events of a decision reaching $i$ (Figure 3.10). The process initiates at $t = 0$ with a distribution concentrated on $X = 0$ (equation 3.24) as a Dirac delta, since it is determined that the subject has no neural activity on $X$.

The equation 3.20 represents the dynamics of the time evolution, where $F_i$ is a force (equation 3.21) deduced from the stochastic equation 3.14. Since the neurophysiological variables $X_i$ are positive, in equation 3.23 a reflective boundary condition must be valid in the set $\Delta = \{X|\exists i, X_i = 0\}$. That considers the perpendicular component of the probability flux on this boundary to be null.

The process evolves while $X$ remains in the interior of the domain $\Omega$ that corresponds to a hypercube of side 1 (Figure 3.10).

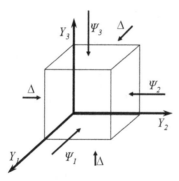

**Fig. 3.10** The Domain $\Omega$ and Boundary Topology $\partial\Omega = \Delta \cup \Psi$ for the Stochastic Process for three Decisions

$P_0(t)$ is the time-dependent probability of not reaching a decision. It can be expanded by the sum over all possible choices of partial probabilities, in which case $P_0(t) = \int_\Omega \phi(X, t)dX$. We can notice that initially at $t = 0$ using the condition 3.24 this probability is $P_0(t = 0) = 1$, which is interpreted as the system starting without a decision. Furthermore, in $t > 0$ the system diffuses and the probability $P_0(t)$ decreases and $\lim_{t\to\infty} P_0(t) = 0$. This property follows directly from the fact that the process has absorbent states on boundary $\Psi$ [71], and the process limit is a stationary state of $\phi(X, t = \infty) = 0$.

The probability $P(t)$ of reaching a decision in time $t$ is the complement of the previous probability $P(t) = 1 - P_0(t)$. The distribution density $p(t)$ is given by the derivative of $P(t)$ as follows in 3.25.

$$p(t) = \frac{\partial P(t)}{\partial t} = -\frac{\partial P_0(t)}{\partial t} = -\int_\Omega \frac{\partial \phi(X, t)}{\partial t} dX = \int_\Omega \nabla \cdot J \, dX = \oiint_{\partial\Omega} J \cdot dS \quad (3.25)$$

$$J_i = \phi F_i - \sigma^2/2 \frac{\partial \phi}{\partial X_i} \quad (3.26)$$

In equation 3.25 uniform convergence is used to distribute the time derivative and the Fokker-Planck equation is used in its continuity form $\frac{\partial \phi}{\partial t} + \nabla \cdot J = 0$ according to the flux expression 3.26. Stoke theorem is used in the integration domain $\Omega$ of figure 3.10. The flux $J \cdot dS$ is interpreted as the probability of crossing an infinitesimal area in time t. Furthermore, the boundary of the $\Omega$ region can be decomposed on several disjoint sets according to $\partial\Omega = (\bigcup_i \Delta_i) \cup (\bigcup_i \Psi_i)$. The surface integral can be separated on each subset.

$$p(t) = \sum_k \oiint_{\Delta_k} J \cdot dS + \sum_k \oiint_{\Psi_k} J \cdot dS \quad (3.27)$$

However, in each plane $\Delta_i$ the orthogonal flux vanishes according to the reflective boundary condition 3.23. Only the term corresponding to the $\Psi_j$ set from equation

3.27 remains. The probability density $p(i, t)$ of making the decision $i$ in time $t$ can be identified by restricting the equation 3.27. In the case that the decision is $i$, all terms with $k \neq i$ vanish, since no flux flows over $\Psi_k$, in which case this probability is given by the total flux over the surface $\Psi_i$.

$$p(i, t) = \oiint_{\Psi_i} J \cdot dS = \int_0^1 \cdots \int_0^1 J_i|_{X_i=1} \prod_{k \neq i} dX_k = -\frac{\sigma^2}{2} \int_0^1 \cdots \int_0^1 \frac{\partial \phi}{\partial X_i}|_{X_i=1} \prod_{k \neq i} dX_k$$

$$(3.28)$$

The expression 3.28 is derived using the border condition $\phi = 0$ on $\Psi$. The minus sign is explained by the fact that if $\phi \geq 0$, then on $\Psi$ the derivatives are negative $\nabla \phi \cdot dS < 0$ so the term is positive. The probability $p(i, t)$ expression is used to construct the maximum likelihood of optimization problems for model calibration.

---

**Algorithm 3.4.1:** WEBUSERSIMULATION($G, p, \kappa, \lambda, \mu$)

---

**comment:** Simple navigational simulation of a single web user visit.

**procedure** DECIDENEXTPAGE($L, \mu, \kappa, \lambda$)

$\begin{cases} \omega \leftarrow \omega(\kappa, \lambda) \\ I_i \leftarrow \frac{e^{\mu \cdot L_i}}{\sum_j e^{\mu \cdot L_j}} \\ X \leftarrow \epsilon \cdot PositiveRandomVector(0, 1) \\ h \leftarrow \epsilon \\ k \leftarrow 0 \\ \textbf{while } (!ReachAdsorbingBoundary(X, \epsilon)) \\ \quad \textbf{do} \begin{cases} k \leftarrow k + 1 \\ Z \leftarrow Gaussian(0, 1) \\ XX \leftarrow E(\omega, t + kh)X(t) + M(\omega, t + kh)I + \sigma K(\omega, t + kh)Z \\ \textbf{if } (ReachReflectiveBoundary(XX, \epsilon)) \\ \quad \textbf{then } X \leftarrow Reflect(XX) \\ \\ \quad \textbf{else } X \leftarrow XX \end{cases} \\ \textbf{return } ((CoordinateReached(X, \epsilon), kh)) \end{cases}$

$sessionList \leftarrow null$
$pp \leftarrow p$
$t \leftarrow 0$
**while** ($pp \neq sink$)
$\quad \textbf{do} \begin{cases} push((pp, t), sessionList) \\ L \leftarrow ExtractLinkTFIDF(pp) \\ (pp, t) \leftarrow DecideNextPage(L, \mu, \kappa, \lambda) \end{cases}$
**return** ($sessionList$)

---

Parameter fitting is a complex task since the higher number of dimension or in this case decision options. Traditional methods for solving differential equation are based on partitioning the space, that solves in a exponential number of step dependent on the dimensionality. A typical web page has 20 hyperlink to decide to click. In [74] a symbolic procedure is proposed for solving this particular differential problem. In [54] a simpler approach based on ant colony technique and hierarchical clustering is used for finding adequate values for the parameter $\mu$. Once the set of parameters is obtained, the system is ready for Monte Carlo simulation. If we considers that parameters of the model are intrinsic to the web users and independent on the web site, then it is plausible to experiment with the effect of small changes in hyperlink structure versus changes in browsing.

A simple Monte Carlo simulation algorithm for web browsing is described in 3.4.1. It depends on the functional form of the matrices $\omega_{ij}(\kappa, \lambda) = \kappa\delta_{ij} + \lambda(1 - \delta_{ij})$, $E(\omega, t) = e^{-\omega t}$, $M(\omega, t) = (1 - e^{\omega t})\omega^{-1}$, and $K(\omega, t) = [\frac{1}{2}(1 - e^{-2\omega t})\omega^{-1}]^{1/2}$. Such matrices can easily be calculated since $\omega$ can be explicitly diagonalized 3.4.1 for any dimension. The algorithm uses random vector generator function like *PositiveRandomVector* that generate uniform distributed vector, and *Gaussian*(*a*, *b*) that generates a Gaussian sampled vector with mean *a* and variance *b*. Border condition are checked with *Reach∗* function within a tolerance given by the global small parameter $\epsilon$. Such a parameter is also used for time step and the starting point. The timed sequences of pages are stored on a linked list *sessionList* using the *push* method.

## 3.5  Discussion

From earlier times static heuristic models of decision making like the Utilitarian Scheme, Random Utility Models, and Expected Utility Theory have been used for describing decision making with considerable success. Nevertheless, axiomatic restrictions (e.g. weak stochastic transitivity) that these models impose are violated in experiments [72], thus constituting paradoxes. The main reason is that people are imperfect information processors and limited in knowledge and processing capabilities.

Time-dependent stochastic processes of decision making were explored in psychology using first principles, thus giving rise to the field of NeuroComputing. Furthermore, models like the Weiner Process, DFT and LCA successfully explain many aspects of the decision making of individuals. Such theories were consistently tested in the laboratory by means of experiments from neuronal activity level to final decision making. However this is still an active research topic, offering promising new proposals like QDT.

On the other side of the spectrum, machine learning pretends to explain everything with complex algorithms trained on observed data. "Wired" magazine published in 2008 [4] an article claiming the end of theories, based on computer power and the information already accumulated. They claimed, "Learning to use a computer of this scale may be challenging. But the opportunity is great; the recent availability of huge amounts of data, along with the statistical tools to crunch these numbers, offers a whole new way of understanding the world. Correlation supersedes causation, and science can advance even without coherent models, unified theories, or really any mechanistic explanation at all. Machine learning could be an efficient answer to theoretical approaches. Nevertheless, machine learning is restricted to the data used for learning." Nothing justifies using the trained algorithm to predict behavior with a variation on the condition of the problem. First principle theories predict changes under such conditions; i.e., they are still the basis for further scientific advancement. As Strogatz [87] paraphrased "full of sound and fury, signifying nothing?", the NetFlix contest is a proof of such non sense. Three years passed without a winner in spite of prolific machine learning research and the one million-dollar reward for improving performance by ten percent.

A dynamic model that describes the navigation decisions a user makes when visiting a web site acquires equal economic importance. In [74] the application of a decision model based on the neurophysiology of decision making simulating the distribution of user visits to a site is successfully described. This model couples stochastic equations to a utility for textual content, which is calibrated using the observed distributions. As a byproduct of this process the distributions of user text preferences are obtained. In addition, web user attention models [66] have been proposed for describing web content importance. Those approaches differ from traditional machine learning by borrowing models from psychology, which are calibrated for explaining behavior instead of recovering behavioral patterns automatically.

Web usage is a very complex system that needs to be studied in detail. The presented theories in this chapter have a long history of scientific research. The hope is to apply them to explain web usage. Both disciples, psychology and Web Intelligence, could be benefited by this convergence. On the one hand, web usage data corresponds to the biggest available repository of human behavior. On the other, first-principle-based models are historically more accurate than phenomenological ones. However, the mathematical complexity is an issue that needs to be further explored and solved.

**Acknowledgements.** This work has been partially supported by the Chilean Millennium Institute of Complex Engineering Systems (ICM: P-05-004-F, CONICYT: FBO16).

# References

1. Abraham, A., Ramos, V.: Web usage mining using artificial ant colony clustering and genetic programming. In: Procs. of the 2003 IEEE Congress on Evolutionary Computation, CEC 2003, pp. 1384–1391 (2003)
2. Akiva, M.B., Lerman, S.: Discrete Choice Analysis: Theory and Application to Travel Demand. MIT Press (1995)
3. Amos, A.: A computational model of information processing in the frontal cortex and basal ganglia. Journal of Cognitive Neuroscience 12(3), 505–519 (2000)

4. Anderson, C.: Wired Magazine, Editorial (June 2008)
5. Anderson, C.R., Domingos, P., Weld, D.S.: Web site personalizers for mobile devices. In: The IJCAI Workshop on Inteligent Techniques for Web Personalization, ITWP 2001 (2001)
6. Anderson, J.R., Bothell, D., Byrne, M.D., Douglass, S., Lebiere, C., Qin, Y.: An integrated theory of the mind. Psychological Review 111(4), 1036–1060 (2004)
7. Audley, R.J., Pike, A.R.: Some alternative stochastic models of choice. British Journal of Mathematical and Statistical Psychology 18, 207–225 (1965)
8. Basso, M.A., Wurtz, R.H.: Modulation of neuronal activity in superior colliculus by changes in target probability. J. Neurosci. 18(18), 7519–7534 (1998)
9. Bhatnagar, V., Gupta, S.K.: Modeling the kdd process. In: Encyclopedia of Data Warehousing and Mining, pp. 1337–1345. IRMA International (2009)
10. Blum, A., Chan, H., Rwebangira, M.R.: A random-surfer web-graph model. In: Proceedings of the Eigth Workshop on Algorithm Engineering and Experiments and the Third Workshop on Analytic Algorithmics and Combinatorics, pp. 238–246. Society for Industrial and Applied Mathematics (2006)
11. Bogacz, R., Brown, E., Moehlis, J., Holmes, P., Cohen, J.D.: The physics of optimal decision making: A formal analysis of models of performance in two-alternative forced choice tasks. Psychological Review 4(113), 700–765 (2006)
12. Borges, J.: A Data Mining Model to Capture User Web Navigation Patterns. PhD thesis, London University (2000)
13. Borooah, V.K.: Logit and probit: ordered and multinomial models, vol. 138 (2001); Quantitative applications in the social sciences. Sage Publications (2002)
14. Britten, K.H., Shadlen, M.N., Newsome, W.T., Movshon, J.A.: Response of neurons in macaque motion signals. Visual Neuroscience 9, 1157–1169 (2006)
15. Buckner, R.L., Kelley, W.M., Petersen, S.E.: Frontal cortex contributes to human memory formation. Nature Neuroscience 2, 311–314 (1999)
16. Busemeyer, J.R., Diederich, A.: Survey of decision field theory. Mathematical Social Sciences 43(3), 345–370 (2002)
17. Busemeyer, J.R., Jessup, R.K., Johnson, J.G., Townsend, J.T.: Building bridges between neural models and complex decision making behaviour. Neural Networks 19(8), 1047–1058 (2006); Neurobiology of Decision Making
18. Busemeyer, J.R., Pothos, E.M., Franco, R.: A quantum theoretical explanation for probability judgment errors. Psychology Revue Letter (2010) (submitted)
19. Busemeyer, J.R., Townsend, J.T.: Decision field theory: a dynamic-cognitive approach to decision making in an uncertain environment. Psychological Review 100(3), 432–459 (1993)
20. Busemeyer, J.R., Wang, Z., Townsend, J.T.: A quantum dynamics of human decision making. Journal of Mathematical Psychology 50(3), 220–241 (2006)
21. Cascetta, E.: Transportation systems engineering: theory and methods. Applied Optimization. Kluwer Academic Publishers (2001)
22. Chanceaux, M., Guérin-Dugué, A., Lemaire, B., Baccino, T.: A Model to Simulate Web Users' Eye Movements. In: Gross, T., Gulliksen, J., Kotzé, P., Oestreicher, L., Palanque, P., Prates, R.O., Winckler, M. (eds.) INTERACT 2009, Part I. LNCS, vol. 5726, pp. 288–300. Springer, Heidelberg (2009)
23. Churchland, A.K., Kiani, R., Shadlen, M.N.: Decision making with multiple choice. Nature Neuroscience 11(6), 693–702 (2008)
24. Cutrell, E., Guan, Z.: What are you looking for?: an eye-tracking study of information usage in web search. In: Proceedings of the SIGCHI Conference on Human Factors in Computing Systems, CHI 2007, pp. 407–416. ACM, New York (2007)

25. de Haan, L., Ferreira, A.: Extreme value theory: an introduction. Springer (2006)
26. de la Cruz Martínez, G., Rodríguez, F.G.: Using user interaction to model user compre-
    hension on the web navigation. International Journal of Computer Information Systems
    and Industrial Management Applications 3, 878–885 (2011)
27. Eliasmith, C.: Computational neuroscience. In: Thagard, P. (ed.) Handbook of Philoso-
    phy of Science, vol. 4, pp. 313–338. Elsevier (2007)
28. Emerson, P.L.: Simple reaction time with markovian evolution of gaussian discriminal
    processes. Psychometrika 35(1), 99–109 (1970)
29. Farzan, R., Brusilovsky, P.: Social Navigation Support for Information Seeking: If You
    Build It, Will They Come? In: Houben, G.-J., McCalla, G., Pianesi, F., Zancanaro, M.
    (eds.) UMAP 2009. LNCS, vol. 5535, pp. 66–77. Springer, Heidelberg (2009)
30. Fasolo, B., McClelland, G.H., Lange, K.A.: The effect of site design and interattribute
    correlations on interactive web-based decisions. In: Online Consumer Psychology: Un-
    derstanding and Influencing Behavior in the Virtual World, pp. 325–344. Psychology
    Press (2005)
31. Glimcher, P.W., Camerer, C., Poldrack, R.A.: Neuroeconomics: Decision Making and
    the Brain. Academic Press (2008)
32. Gold, J., Shadlen, M.: The neural basis of decision making. Annual Review of Neuro-
    science 30, 535–574 (2007)
33. Google. Investor relation: Finacial table (2009),
    http://investor.google.com/financial/tables.html
34. Grossberg, S., Gutowski, W.E.: Neural dynamics of decision making under risk: Affec-
    tive balance and cognitive-emotional interactions. Psychological Review 94(3), 300–
    318 (1987)
35. Hanks, T.D., Ditterich, J., Shadlen, M.N.: Microstimulation of macaque area lip affects
    decision-making in a motion discrimination task. Nature Neuroscience 9(5), 682–689
    (2006)
36. Hebb, D.: The organization of behaviour: a neuropsychological theory. L. Erlbaum As-
    sociates (1949)
37. Heekeren, H.R., Marrett, S., Ungerleider, L.G.: The neural systems that mediate human
    perceptual decision making. Nature Reviews Neuroscience 9, 467–479 (2008)
38. Helander, M.G., Khalid, H.M.: Modeling the customer in electronic commerce. Applied
    Ergonomics 31, 609–619 (2000)
39. Hill, A.V.: Excitation and accommodation in nerve. Proceedings of the Royal Society
    B 119, 305–355 (1936)
40. Huberman, B.A., Pirolli, P.L.T., Pitkow, J.E., Lukose, R.M.: Strong regularities in world
    wide web surfing. Science 280(5360), 95–97 (1998)
41. Jimenez-Molina, A., Ko, I.-Y.: Cognitive resource aware service provisioning. In: The
    2011 IEEE / WIC / ACM International Conference (2011) (to appear)
42. Kahneman, D.: Maps of Bounded Rationality:a perspective on intuitive judgment and
    choice. In: The Nobel Prizes 2002. Nobel Foundation (2003)
43. Kahneman, D., Tversky, A.: Prospect theory: An analysis of decision under risk. Econo-
    metrica 47, 263–291 (1979)
44. Karampatziakis, N., Paliouras, G., Pierrakos, D., Stamatopoulos, P.: Navigation Pattern
    Discovery Using Grammatical Inference. In: Paliouras, G., Sakakibara, Y. (eds.) ICGI
    2004. LNCS (LNAI), vol. 3264, pp. 187–198. Springer, Heidelberg (2004)
45. Karanam, S., Van Oostendorp, H., Indurkhya, B.: The role of content in addition to hy-
    perlinks in user-clicking behavior. In: Proceedings of the 28th Annual European Con-
    ference on Cognitive Ergonomics, ECCE 2010, pp. 125–131. ACM, New York (2010)

46. Kiani, R., Shadlen, M.N.: Representation of confidence associated with a decision by neurons in the parietal cortex. Science 324(8), 759–764 (2008)
47. Kim, J.-N., Shadlen, M.N.: Neural correlates of a decision in the dorsolateral prefrontal cortex of the macaque. Nature Neuroscience 2(2), 176–185 (1999)
48. Kitajima, M., Polson, P.G., Blackmon, M.H.: Colides and snifact: Complementary models for searching and sensemaking on the web. In: Human Computer Interaction Consortium (HCIC) Winter Workshop (2007)
49. Kitajima, M., Blackmon, M.H., Polson, P.G.: Cognitive Architecture for Website Design and Usability Evaluation: Comprehension and Information Scent in Performing by Exploration. In: Proceedings of the Human Computer Interaction International Conference (2005)
50. Korfiatis, G., Paliouras, G.: Modeling web navigation using grammatical inference. Applied Artificial Intelligence 22(1&2), 116–138 (2008)
51. Laberge, D.: A recruitment theory of simple behavior. Psychometrika 27(4), 375–396 (1979)
52. Laming, D.R.J.: Information theory of choice reaction time. Wiley (1968)
53. Lohr, S.: A 1 million dollars research bargain for netflix, and maybe a model for others. New York Times (2009)
54. Loyola, P., Román, P.E., Velásquez, J.D.: Clustering-based learning approach for ant colony optimization model to simulate web user behavior. In: 2011 IEEE / WIC / ACM International Conference, France (2011)
55. Luna, R., Hernandez, A., Broddy, C.D., Romo, R.: Neural codes for perceptual discrimination in primary somatosensory cortex. Nature Neuroscience 8(9), 1210–1219 (2005)
56. Manning, C.D., Schutze, H.: Fundation of Statistical Natural Language Processing. The MIT Press (1999)
57. Mas-Colell, A., Whinston, M.D., Green, J.R.: Microeconomic Theory. Oxford University Press (1985)
58. Mcclelland, J.L.: Toward a theory of information processing in graded, random, and interactive network. In: Meyer, D.E., Kornblum (eds.) Attention and Performance XIV: Synergies in Experimental Psychology, Artificial Intelligence, and Cognitive Neuroscience, pp. 665–688. MIT Press (1993)
59. McFadden, D.: Is conditional logit analysis of qualitative choice behavior. In: Zarembkaá (ed.) Frontiers in Econometrics. Academic Press (1973)
60. Meyer, D.E., Irwin, D.E., Osman, A.M., Kounios, J.: The dynamics of cognition and action: mental processes inferred from speed-accuracy decomposition. Psychol. Rev. 95(2), 183–237 (1988)
61. Miller, C.S., Remington, R.W.: Modeling information navigation: implications for information architecture. Human Computer Interaction 19(3), 225–271 (2004)
62. Navon, D.: On the economy of the human-processing system. Psychological Review 86(3), 214–255 (1979)
63. O'Reilly, R.C.: The what and how of prefrontal cortical organization. Trends in Neuroscience 33(8), 355–361 (2010)
64. Philiastides, M.G., Heekeren, H.R.: Spatiotemporal characteristics of perceptual decision making in the human brain. In: Dreher, D.J.-C., Tremblay, L. (eds.) Handbook of Reward and Decision Making, pp. 185–212. Academic Press, New York (2009)
65. Pirolli, P.L.T.: Information Foraging Theory: Adaptive Interaction with Information. Oxford University Press (2007)
66. Pirolli, P.L.T.: Power of 10: Modeling complex information-seeking systems at multiple scales. Computer 42, 33–40 (2009)

67. Pirolli, P.L.T., Fu, W.-T.: Snifact: a model of information foraging on the world wide web. In: 9th International Conference on User Modeling (2003)
68. Pothos, E.M., Busemeyer, J.R.: A quantum probability explanation for violation of rational decision theory. Proceedings of The Royal Society B 276(1165), 2171–2178 (2009)
69. Ratcliff, R.: A theory of memory retrieval. Psychological Review 85(2), 59–108 (1978)
70. Ratcliff, R., Van Zandt, T., McKoon, G.: Connectionist and diffusion models of reaction time. Psychological Revue. 106(2), 261–300 (1999)
71. Resnick, S.I.: Adventures in stochastic processes. Birkhauser Verlag, Basel (1992)
72. Rieskamp, J., Busemeyer, J.R., Mellers, B.A.: Extending the bounds of rationality: Evidence and theories of preferential choice. Journal of Economic Literature 44(3), 631–661 (2006)
73. Roitman, J.D., Shadlen, M.N.: Response of neurons in the lateral intraparietal area during a combined visual discrimination reaction time task. Journal of Neuroscience 22, 9475–9489 (2002)
74. Román, P.E.: Web User Behavior Analysis. PhD thesis, University of Chile (January 2011)
75. Román, P.E., Velásquez, J.D.: Analysis of the web user behavior with a psychologically-based diffusion model. In: Of the AAAI, Fall Symposium on Biologically Inspired Cognitive Architectures, Arlington, USA., Arlington, Washington DC, USA, Technical Paper of the AAAI (2009)
76. Román, P.E., Velásquez, J.D.: A dynamic stochastic model applied to the analysis of the web user behavior. In: Snasel, et al. (eds.) The 2009 AWIC 6th Atlantic Web Intelligence Conference, Prague, Czech Republic. Invited Lecture. Intelligent and Soft Computing Series, Advances in Intelligent Web Mastering-2, pp. 31–40 (2009)
77. Román, P.E., Velásquez, J.D.: Artificial web user simulation and web usage mining. In: The First Workshop in Business Analytics and Optimizatio áBAO 2010, Santiago, Chile (January 2010)
78. Román, P.E., Velásquez, J.D.: Stochastic simulation of web users. In: Procs. of the 2010 IEEE / WIC / ACM International Conference, Toronto, Canada. IEEE Press (September 2010)
79. Román, P.E., Velásquez, J.D.: The time course of the web user. In: Second Workshop on Time Use Observatory áTUO2, San Felipe, Chile (March 2010)
80. Rubinstein, R.Y., Kroese, D.P.: Simulation and the Monte Carlo method. Wiley series in probability and mathematical statistics. Probability and mathematical statistics. John Wiley & Sons (2008)
81. Sato, T., Murthy, A., Thompson, K.G., Schall, J.D.: Search efficiency but not response interference affects visual selection in frontal eye field. Neuron 30, 583–591 (2001)
82. Schall, J.D.: Neural basis of deciding, choosing and acting. National Review of Neuroscience 2(1), 33–42 (2001)
83. Schall, J.D.: On building a bridge between brain and behavior. Annual Review of Psychology 55, 23–50 (2004)
84. Schall, J.D.: Frontal eye fields. In: Encyclopedia of Neuroscience, vol. 4, pp. 367–374. Elsevier (2009)
85. Scott, M.L.: Programming language pragmatics. Morgan Kaufmann Publishers Inc., San Francisco (2000)
86. Stone, M.: Models for choice reaction time. Psychometrika 25(3), 251–260 (1960)
87. Strogatz, S.H.: Romanesque networks. Nature 433(27), 365–366 (2005)
88. Michael Jahrer, A.T., Bell, R.M., Park, F.: The bigchaos solution to the netflix grand prize. NetFlix, 1–52 (2009)

89. Telang, R., Kumar, A.: Impact of customer web portals on call center: An empirical analysis. SSRN eLibrary (2009)
90. Train, K.: Discrete choice methods with simulation. Cambridge University Press (2009)
91. Tversky, A., Kahneman, D.: The framing of decisions and the psychology of choice. Science 211(4481), 453–458 (1981)
92. Tversky, A., Sattath, S.: Preference trees. Psychological Review 86(6), 542–573 (1979)
93. Tversky, A., Simonson, I.: Context-dependent preferences. Management Science 39(10), 1179–1189 (1993)
94. Usher, M., McClelland, J.L.: The time course of perceptual choice: The leaky, competing accumulator model. Psychological Review 2(1), 550–592 (2001)
95. Velasquez, J.D., Palade, V.: A knowledge base for the maintenance of knowledge. Journal of Knowledge Based Systems 1(20), 238–248 (2007)
96. Van Neumann, J., Morgenstern, O.: Theory of Games and Economic Behavior (Commemorative Edition) (Princeton Classic Editions), 60 anv edn. Princeton University Press (2007)
97. Wald, A., Wolfowitz, J.: Optimum character of the sequential probability ratio test. The Annals of Mathematical Statistics 19(3), 326–339 (1948)
98. Wickens, C.D.: Multiple resources and performance prediction. Theoretical Issues in Ergonomics Science 3(2), 159–177 (2002)
99. Wickens, C.D.: Multiple Resources and Mental Workload. Hum. Factors 50(3), 449–455 (2008)
100. Wickens, T.D.: Elementary Signal Detection Theory. Oxford University Press (2002)
101. Zeng, L.: A heteroscedastic generalized extreme value discrete choice model. Sociological Methods Research 29(1), 118–144 (2000)

# Chapter 4
# Web Usage Mining: Discovering Usage Patterns for Web Applications

Giovanna Castellano, Anna M. Fanelli, and Maria A. Torsello

**Abstract.** The heterogeneous nature of the Web combined with the rapid diffusion of Web-based applications have made Web browsing an intricate activity for users. This has given rise to an urgent need for developing systems capable to assist and guide users during their navigational activity in the Web. Web Usage Mining (WUM) refers to the application of Data Mining techniques for the automatic discovery of meaningful usage patterns characterizing the browsing behavior of users, starting from access data collected from interactions of users with sites. The discovered patterns may be conveniently exploited in order to implement functionalities offering useful assistance to users. This chapter is mainly intended to provide an overview of the different stages involved in a general WUM process. As an example, a WUM approach is presented which is based on the use of fuzzy clustering to discovery user categories starting from usage patterns.

## 4.1 Introduction

Nowadays due to the significant increase of data available in Internet combined with its rapid and disordered growth, the World Wide Web has evolved into a network of data without a proper organizational structure. In addition, the variegated and heterogeneous nature of the network have made Web browsing an intricate activity not only for inexperienced users but also for expert users. As a consequence, when browsing the Web users often feel lost, overwhelmed by a huge quantity of data that continue to enlarge over time. Moreover e-business and web marketing are quickly developing, offering a big variety of services. Thus making predictions on the needs of customers has become a crucial activity of many Web applications as well as

Giovanna Castellano · Anna M. Fanelli · Maria A. Torsello
University of Bari "A. Moro", Department of Informatics,
Via Orabona, 4 - 70125 Bari, Italy
e-mail: {castellano,fanelli,torsello}@di.uniba.it

J.D. Velásquez et al. (Eds.): Advanced Techniques in Web Intelligence-2, SCI 452, pp. 75–104.
springerlink.com              © Springer-Verlag Berlin Heidelberg 2013

understanding or anticipating the preferences of users with the aim to improve the usability of Web and, at the same time, to meet the user needs.

With the proliferation of Web-based systems, massive amounts of user data collected during their interactions with applications are generated. The analysis of such data could be useful to achieve several goals strictly related to the nature of the considered Web applications. These include providing personalized content or services to users, designing marketing strategies, optimizing functionalities of Web applications, etc. The analysis of data collected by user interactions with Web applications is essentially aimed at the automatic discovery of meaningful patterns from large collections of Web data.

Generally speaking, the adoption of machine learning techniques reveals to be an appropriate way to analyze data collected on the Web and extract useful knowledge starting from these. The efforts carried out in this direction have led to the growth of an interesting research area named *Web mining* [47, 6, 30] that essentially refers to the application of Data Mining methods in order to automatically discover and extract knowledge from data generated by the Web. Commonly, according to the different types of Web data which are exploited in the process of knowledge discovery, three main areas can be distinguished in Web mining. These are *Web content mining*, *Web structure mining*, and *Web usage mining*.

Web content mining [52, 21] concerns the discovery of useful information from the contents of Web documents. In general, Web content could encompass a very broad range of data, such as text, images, audio, video, metadata as well as hyperlinks, or structured records such as lists and tables. Recently research in this field is focusing on mining multi types of data, leading to a new branch called *multimedia data mining* representing a particular instance of the Web content mining.

Web structure mining [25, 29] tries to discover the model underlying the link structures of the Web. The model is typically based on the topology of the hyperlinks characterizing the structure of the Web graph. The discovered models are usually used to categorize Web pages and to generate information about the relationships or the similarity degrees existing among different Web pages.

Web Usage Mining (WUM) [70, 55, 27, 91] aims at discovering interesting patterns by exploiting usage data stored during the interactions of users with the Web site that generally characterize the navigational behavior of users. Web usage data mainly consist of huge collections of data deriving from several sources such as Web server access logs, proxy server logs, registration form data, mouse clicks, and any other source that allow to collect data representing the result of user interactions. Broadly speaking WUM provides an overall approach to the collection and preprocessing of usage data, and the extraction of patterns encoding the behavior and the preferences of users. The discovered patterns are usually represented as collections of pages, objects, or resources that are frequently accessed by an individual user or, more frequently, by groups of users having the same needs or interests.

In this work we mainly focus on the WUM area. In recent years, a large number of works have been published that illustrate the state-of-the-art and the advances in the field of WUM [8, 36]. The results in this research field have become critical for a wide number of applications such as business and marketing decision support, usability studies, network traffic analysis, etc. To mention few examples, knowledge derived from Web usage patterns could be directly applied to efficiently manage activities related to e-business, e-services, e-education and so on [1, 18, 20]. In e-business, for example, the analysis of usage information from the Web could be helpful to attract new customers, retain current customers, improve cross marketing/sales, increase the effectiveness of promotional campaigns, etc. [37, 40]. Other typical applications in WUM are those that take advantage from different user modeling techniques such as the design of adaptive Web sites and recommender systems. Actually, WUM represents one of the most employed approaches for the development of Web personalization systems, as also demonstrated by the large number of research works published on this topic [1, 70, 27, 57].

The main objective of this chapter is to provide a comprehensive overview of the major issues related to the task of mining usage patterns modeling the browsing behavior of Web users. In addition, we intend to discuss the main applications that may benefit from the adoption of Web usage-based approaches for the achievement of their ultimate goals. Moreover, the work illustrates an application example of a clustering technique for the discovery of user models with the aim to emphasize on the suitability and effectiveness of clustering techniques in the pattern discovery process.

The rest of the chapter is organized as follows. In section 4.2, we describe the different stages involved in a generic WUM process, emphasizing on the aspects and the problems characterizing each stage along with the different techniques employed. Section 4.3 presents a WUM-based approach for the discovery of user categories describing the common characteristics of groups of users having the same interests. The presented approach is based on the employment of a fuzzy relational clustering algorithm embedding a fuzzy measure for the evaluation of the similarity among Web users. To close the work, section 4.4 reports some conclusive remarks.

## 4.2   The WUM Process

According to a standard data mining process, four main inter-dependent stages can be recognized in a general WUM process [57]:

- **Web usage data collection.** Web usage data are gathered from various sources by using different techniques that allow to attain efficient collections of user data.
- **Web usage data preprocessing.** The collected Web usage data are preprocessed to obtain data expressed in a form that is suitable to be analyzed in the next steps. Specifically, in this stage, data are cleaned from noise, inconsistencies are solved and finally data are organized in an integrated and consolidated manner.
- **Web usage pattern discovery.** The available usage data are analyzed in order to mine significant correlations between data and discover usage patterns, reflecting the behavior and the interests of users. Typically, in this stage learning methods, such as clustering, association rule discovery and sequential pattern discovery are applied in order to automate the process of pattern discovery.

- **Application of Web usage patterns.** The extracted knowledge expressed in terms of usage patterns is analyzed and evaluated in order to be exploited to implement the effective functionalities of the considered application such as recommendation engines, visualization tools, Web analytics and report generation tools.

In the following we provide a detailed description of each stage along with the proposed techniques.

## 4.2.1  Web Usage Data Collection

As in any data mining application, a crucial task to be performed in WUM is the creation of an appropriate collection of relevant Web data which will be analyzed through the application of data mining and statistical techniques in order to provide useful information about the user navigational behavior. Data preparation is a time consuming and computationally intensive step in a WUM process, requiring the adoption of special algorithms and heuristics not commonly employed in other domains. This is essentially due to the intrinsic characteristics of usage data that can be collected from various kinds of data sources. In the following we provide a summary of data types that are commonly employed in WUM processes.

**Server Side Data.** Web servers represent the richest and the most common source of Web data because they explicitly record large amounts of information characterizing the browsing behavior of site visitors.

Data collected at the server side principally include several types of log files generated by the Web server. Data recorded into the server log files reflect the (eventually concurrent) accesses to a Web site by multiple users in chronological order. These log files can be stored in various formats. Most of the Web servers support as a default option the Common Log File format (CLF) which typically includes information such as the IP address of the connected user, the time stamp of the request (date and time of the access), the URL of the requested page, the request protocol, a code indicating the status of the request, the size of the page (if the request is successful). Other examples of formats of log files are represented by the Extended Log Format (W3C) supported by Web servers as Apache and Netscape, and the very similar W3SVC format supported by Microsoft Internet Information Server. Such formats are characterized by the inclusion of additional information about the user requests, like the address of the referring URL to the requested page, the name and the version of the browser used by the user for the navigation, the operating system of the host machine.

Data recorded in log files may not be always entirely reliable. The problem of the unreliability of these sources of data is mainly due to the presence of various levels of caching within the Web environment and to the misinterpretation of the IP user addresses.

The requests for cached Web pages are not recorded into log files. In fact, whenever a user accesses to the same Web page, rather than making a new request to the server, the cached copy is returned to the user. In this way, the user request does not reach the Web server holding the page and, as a result, the server is not aware of the actions and the page accesses made by the users. Cache-busting represents one solution to this first problem. This involves the use of special headers, defined either in Web servers or Web pages, that include directives to establish the objects that should be cached, and the time that they should be cached.

The problem of IP address misinterpretation is essentially caused by two reasons. With the use of the intermediate proxy server which assigns to all users the same IP address, the requests from different host machines passing through the proxy server are recorded into log files with the same IP. The same problem occurs when different users use the same host machine. The dynamic IP allocation gives raise to the opposite situation, where different addresses may be assigned to the same user. Both these problems may cause serious complications, especially in those WUM processes where the identification of individual users is fundamental.

Other kinds of usage data can be stored by the Web server through the dispensation and tracking of cookies. Through the cookie mechanism the Web server can store its own information about the user in a file within the client machine. This information (usually consisting in a unique ID) will be used by the server to recognize the user during his/her successive accesses to the site. The use of cookies has raised important concerns about the user privacy and security. Indeed, this mechanism requires the cooperation of users that for different reasons could choose to disable the option of accepting cookies.

Another kind of data which the Web server can collect is represented by data explicitly supplied by users during their interactions with the site. This kind of data are typically obtained through the fulfillment of apposite registration forms which provide important demographic and personal information or also explicit knowledge about the user preferences. However these data are not always reliable, since users often provide incomplete and inaccurate information. Additional explicit user data collected at the server side consist in the query data generated by online visitors while searching for pages relevant to their information needs [12].

*Client Side Data.* Usage data collected at the client side are essentially represented by data originated by the host accessing the Web site.

A first way to collect client side data is through remote agents (generally implemented in Java or Javascripts) which are embedded in Web pages, such as Java applets [79]. Such agents allow to directly collect information from the client such as the user browsing history, the pages visited before visiting the current page, the sites visited before and after the current site, the time that the user accesses to the site and when he leaves it. This mechanism permits to collect more reliable data since it is able to overcome the previously mentioned limitations of Web cache and IP misinterpretation underlying the adoption of server log files. However such mechanism requires users cooperation in enabling the functionality of the Javascript and Java applets on their machines. In fact, since the employment of remote agents may affect the client system performances by introducing additional overhead whenever the users try to access the Web site, users may choose to disable these functionalities on their systems.

An older mechanism used to collect usage data from the client host consists in modifying the source code of an existing browser to enhance its capabilities of data collection. Based on this mechanism browsers are modified in order to enable the storing of information about the user navigational behavior, such as the Web pages visited by users, the access time, the response time of the server, etc. As for the use of remote agents the user cooperation is necessary. Modified versions of browsers are often considered a threat to the user privacy. Thus one main difficulty inherent this method of data collection consists in convincing users to use these modified browser versions. A way often used to overcome this difficulty consists in offering incentives to users such as additional software or services. Anyway, modifying a modern browser is not a simple task, even when its source is available.

**Intermediary Data.** Another important source of data reflecting the user browsing behavior is represented by the proxy server, a software system which plays the role of intermediary between the client browser and the Web server ensuring security, administrative control and caching services. Proxy caching represents a way to reduce the loading time of a Web page as well as the network traffic load at the server and client sides [22]. This intermediate employs logs having similar format to server log files. This is the main advantage of using these logs. In fact, since proxy caching reveals the requests of multiple clients to multiple servers, this can be considered a valuable source of data characterizing the navigational behavior of a group of anonymous users sharing a common proxy server [83].

Packet sniffers provide an alternative method of usage intermediary data collection. A packet sniffer is a piece of software (sometimes a hardware device) which is able to monitor the network traffic coming to a Web server and to extract usage data directly from TCP/IP packets. On the one hand, the use of packet sniffers has the advantage that data are collected and analyzed in real time. On the other hand, since data are not logged, this can give raise to a strong drawback: data can be lost when, for example, something goes wrong either with the packet sniffer or with the data transmission.

## 4.2.2   Web Usage Data Preprocessing

The second stage to be performed in any WUM process is the preprocessing of Web usage data. Usage data collected from various sources (as specified above) are usually bulky and characterized by noise, ambiguity and incompleteness. Thus, they need to be assembled in order to obtain data collections expressed in a consistent and integrated manner to be used as input to the next step of pattern discovery. To accomplish this a preliminary activity of data preprocessing reveals to be necessary. In general data preprocessing involves the execution of a set of operations such as the elimination of noise, the solution of inconsistencies, the fulfillment of eventual missing values, the removal of redundant or irrelevant data. Data preprocessing is highly dependent on the problem domain and the quality and type of available data. Hence, this stage requires an accurate analysis of data and represents one of the hardest task in a general WUM process. An important facet to be considered consists in the trade-off regarding the preprocessing stage. On the one hand, an insufficient preprocessing could make more difficult the successive task of pattern discovery. On the other hand, an excessive preprocessing could remove data embedding implicit knowledge that could be useful for the next steps of the WUM process. As a consequence, the correct application of data preprocessing tasks strongly affects the success of the overall WUM process. An extensive description of data preparation and preprocessing methods can be found in [23, 39, 76, 19].

The main goal of data preprocessing is to transform and to aggregate the raw data into different levels of abstraction which can be useful for achieving the specific aims of the overall WUM process. To obtain significant data abstractions, three main activities are performed in data preprocessing stage, namely *data cleaning, user identification* and *user session identification*. In the sequel a description of these activities is provided, with a focus on the methods adopted to carry out the respective tasks.

## Data Cleaning

Data cleaning represents a fundamental activity in Web usage data preprocessing devoted to clean raw Web data from noise. This activity mainly concerns server side data being these data particularly affected by noise. Hence, the rest of the discussion about data cleaning will focus on access log files.

Since Web log files record all the interactions between Web site and its users, they may include a lot of useless information. One first goal of data cleaning consists in removing from log files data fields that do not provide significant information in analysis or data mining tasks (for example the number of bytes transferred, the version of the used HTTP protocol, etc.). One main aim of data cleaning is to remove from log files those records corresponding to irrelevant and redundant requests. Redundant records in log files are mainly due to the fact that a separate access request is generated for every file, image, and multimedia object embedded in the Web page requested by the user. In this way a single user request for a Web page may often result in several log entries that correspond to files automatically downloaded without an explicit request of the same user. Since these records do not represent the effective browser activity of the connected user, they are deemed redundant and they have to be deleted. One simple way to eliminate these items consists in checking the suffix of the URL name and, for example, removing all log entries with filename suffixes such as gif, jpeg, jpg, etc. The list of the suffixes can be properly modified according to the type of the considered Web application. For example, for an application mainly characterized of multimedia contents, the elimination of the requests to the previous type of files could cause the loss of relevant information [24].

Irrelevant records that are removed correspond also to failed user requests containing for example an error status code.

Another crucial task of data cleaning concerns the identification and elimination of accesses generated by Web robots (also known as Web crawlers or Web spiders). These are programs which traverse the Web in a methodical and automated manner, downloading complete Web sites in order to update the index of a search engine. The entries generated by Web robot are filtered out from the log files since they are not considered as usage data representative of the actual user browser behavior. In conventional techniques, Web robot sessions are detected in different ways: by examining sessions that access a specially formatted file called robots.txt, by exploiting the User Agent field of log files wherein most crawlers identify themselves, or by matching the IP address of sessions with those of known robot clients. A robust technique to detect spider sessions is based on the assumption that the behavior of robots is different from those of human users. Web robots are recognized by using a set of relevant features extracted from access logs (percentage of media files requested, percentage of requests made by HTTP methods, average time between requests, etc.) [86]. A simpler method to recognize robots is to monitor the navigational behavior pattern of the user. Specifically, if a user accesses to all links of all the pages of a Web site, it will be considered a crawler.

**User Identification**

One of the most complex activities in Web usage data preprocessing is represented by user identification. Indeed the task of identifying a single user is crucial in order to characterize his browsing behavior.

A large number of Web applications recognize the user by means of the explicit request of registration by filling a special form. However this way of recognizing users presents some drawbacks. The user might be reluctant to share personal information and the burden of registration could discourage the navigation of the site. As a consequence, different approaches able to automatically identify users have been proposed in literature.

Among all the approaches proposed, the simplest and also the mostly adopted approach consists in considering each different IP address included in log files as a single user [64, 85]. However, this approach is not very accurate since a user may access the Web from different computers or many users may use the same IP address (if a proxy is used) to access the Web.

Other WUM tools use cookies for the identification of unique users [43]. Also the employment of cookies is not without problems mainly due to the possibility for users to disable cookies on their systems.

An alternative approach proposed in [72] for user identification consists in the use of special Internet services, such as the *inetd* and *fingerd*, which provide the name and other information about the user accessing the Web server. However, as for cookies, also these services can be disabled by users. To overcome this limitation, further approaches have been developed.

In [23], two different heuristics have been proposed to identify users. Based on the first method, Web log files (expressed in the Extended Log Format) are analyzed by searching for different browsers or different operating systems, even when the IP address is the same. This suggests that the requests are originated from different users. Conversely, in the second method, the topology of the Web site is exploited to recognize requests generated from different users. More precisely, a new user is recognized whenever a request for a Web page derives from the same IP address of the requests for other Web pages but no link exists between these pages.

**User Session Identification**

In a general WUM process usage data are properly analyzed in order to discover the user browsing behavior during his/her interactions with Web applications. This knowledge is generally embedded in user sessions. Of course the identification of user sessions constitutes another important activity in Web usage data preprocessing that strongly affects the quality of the usage patterns discovered in the next stage of the WUM process.

According to [85], a user session can be defined as a delimited set of URLs corresponding to the pages accessed by the same user within a particular visit to a Web site. Hence, once a user has been recognized, the next step of Web usage data preprocessing is to identify the respective user session, namely the clickstream of each user is portioned into logical structures that are sessions.

As in [81], the existing methods for user session identification can be divided into two main categories: time-oriented and navigation-oriented approaches.

In time-oriented approaches [28, 23, 80] two information are usually taken into account: the total session time and the single page-stay time. In the first case, consecutive accesses made within a minimum fixed timeout are considered belonging to the same session. In the other case, two consecutive accesses that exceed a maximum fixed timeout are considered to belong to different sessions. Different values have been chosen for setting these thresholds depending on the content of the considered application and the particular goal of the WUM process. Based on empirical studies, many commercial products established for the total session time threshold values varying from 30 to 25.5 minutes while for the page-stay time, a default value of 10 minutes has been adopted in many works [19]. Nevertheless, time-based approaches are not very reliable since users may be involved in other activities when browsing the Web and, in addition, factors such as busy communication line, loading time of components in web page and content size of Web pages are not considered in this kind of approach.

Navigation-oriented approaches substantially exploit the Web application topology in graph format to identify user sessions. In this approach, transactions are recognized where a transaction represents a subset of pages that occur in a user session. Starting from the assumption that transactions depend on the contextual information, Web pages are classified in: auxiliary, content and hybrid pages. Auxiliary pages contain links to other pages of the site; content pages contain the information interesting for the user and the hybrid pages are of both previous kinds of pages. Starting from this classification, in [23] it has been possible to distinguish content-only transactions from the auxiliary-content transactions. The first ones include all the content pages visited by the user whereas the second ones refer to the paths to retrieve a content page.

Although the session identification approaches mentioned above are not innovative, most of the recent applications [10, 31, 65, 32, 44] adopt them to derive sessions. To increase the performance of the user session identification process, different methods combining both kinds of approaches have been proposed in several literature works. For example in [31], firstly a navigation-oriented heuristic is employed to identify sessions. Specifically, the Referrer URL field is checked and a new user session is identified if the URL in the Referrer URL field has never been accessed before, or there is a large interval between the access time of this record and the previous one if the Referrer URL field is empty. Then, if the identified sessions include more than one visit by the same user at different time, a time-oriented heuristic is used to segment the different visits into different user sessions. A novel session construction method, called Smart-SRA, has been proposed in [7]. This method models session construction process as a graph problem and produces maximal paths traversed on the web graph. In Smart-SRA server request log sequences are processed to reconstruct a user session as a set of valid navigation paths. This is accomplished by means of two phases. In the first phase, the access log data of Web users are partitioned into shorter page request sequences called candidate sessions, by using session duration time and page-stay time rules. In the second phase, candidate sessions are divided into maximal sub-sessions such that for each consecutive page pair in the sequence there exists a link from previous one to latter one.

### 4.2.3   Web Usage Pattern Discovery

Once Web usage data have been preprocessed, the next stage of the WUM process involves the discovery of usage patterns that can be properly exploited to realize the functionalities of the specific Web application. This aim is pursued by applying a variety of methods and algorithms falling into several research fields such as statistics, data mining, machine learning and pattern recognition able to discover from usage data useful knowledge about the browsing behavior of users.

Most of commercial applications commonly mine knowledge about user behavior through the application of standard statistical techniques on session data. As a result, many WUM traffic tools provide periodic reports summarizing important statistical information descriptive of the user browser patterns, such as most frequently accessed pages, average view time, average length of navigational paths, etc. The resulting knowledge may be exploited to achieve different goals such as to improve the system performance, to facilitate the site modification task, to provide support for marketing decisions, etc. There are several commercial software that could provide Web usage statistics (Analog, 2003; ClickTracks, 2003; Hitbox, 2003; LogRover, 2003; Website Tracker, 2003; WebStat, 2003). For small web servers, the usage statistics provided by conventional Web site trackers may be adequate to analyze the usage pattern and trends. However as the size and complexity of the data increases, the statistics provided by existing Web log file analysis tools may prove inadequate and more intelligent knowledge mining techniques will be necessary [71, 74, 93, 2, 90].

In the context of knowledge discovery techniques specifically designed for the analysis of Web usage data, research efforts have been mainly addressed to the adoption of four main distinct paradigms: association rules, sequential patterns, clustering and classification. Exhaustive reviews of the works focusing on these techniques are given in [33, 57]. In the following we discuss the most common pattern discovery techniques employed in the WUM domain together with some application examples.

## Association Rule Mining

The most straightforward technique for pattern discovery in WUM consists in mining association rules that essentially try to find significant associations among groups of pages which frequently appear in user sessions.

Most common approaches to association rule discovery are based on Apriori, an algorithm (detailed in [3]) that is able to detect frequent itemsets, i.e. items that occurr frequently together in many transactions. Then, starting from the frequent itemsets, association rules which satisfy a minimum confidence threshold are generated.

Generally speaking, an association rule can be expressed in the following form:

$$X \Rightarrow (Y - X)$$

where $X$ and $Y$ are frequent itemsets and $X \subset Y$. Itemsets $X$ and $(Y - X)$ respectively represent the antecedent and the consequence of the rule. The generated association rules are valid only if the measure of confidence results to be greater or equal than an established minimal threshold. An exhaustive description of the association rule mining technique can be found in [3, 4].

The process of association rule discovery can be particularly profitable in e-commerce context by enabling the analysis of how items purchased by customers in a on-line shop are related. An example of employment of association rules in a top-N recommender system for e-commerce is presented in [75]. In this work, preferences of a user are matched against the items in the antecedent of each discovered rule, and the items on the consequent part of the matching rules are sorted in a list according to the confidence values. Finally the top-N ranked items from this list are recommended to the user. Association rules can also be used with the aim of optimizing the structure of a site. For example, if in a site no direct link exists between two pages *A* and *B*, the discovery of a rule $A \Rightarrow B$ would mean that adding the link from *A* to *B* could assist users in finding more quickly the desired information.

Association rule mining technique has been used in [42, 60]. Some measures of interest to evaluate association rules mined from Web usage data have been proposed by [38]. Fuzzy association rules obtained by the combination of association rules and fuzzy logic have been extracted in [88].

A main disadvantage of association rule mining is that the information about the temporal sequence is not considered. As a consequence, the discovered patterns relate items regardless of the order in which they occur in a given transaction.

## Sequential Pattern Discovery

Sequential pattern discovery turns out to be particularly useful for the identification of navigational patterns in Web usage data. In this approach, time is introduced in the process of discovering patterns. The attempt of this technique is to discover time ordered sequences of pages that appear frequently in user sessions. In general a typical sequential pattern is expressed in a form syntactically similar to association rules. As an example, a sequential pattern can be the following: the 80% of users who firstly visited the page A.html and then visited the page B.html, they accessed also the page C.html within the same session.

To extract sequential patterns methods based on association rule mining and methods based on the use of tree structures and Markov chains are commonly used. A comparison of different sequential pattern algorithms applied to WUM is presented in [58].

Some well-known algorithms for mining association rules have been modified to obtain sequential patterns. For example, the Apriori algorithm has been properly extended to derive two new algorithms, namely the AprioriAll [38] and GSP [58].

An alternative approach involves the use of Markov models to model the navigational activity of users. According to this approach, each page can be represented as a state and the transition probability between two states represent the probability that the user navigates from one state to another. The main goal of such approaches is to predict the next action of the user based on his/her navigational behavior history. A mixture of Markov models has been adopted in [13]. Here, different Markov models are used for different clusters in order to characterize and visualize the navigational behavior of various user categories. A very similar method was used in [5] to individuate sequential patterns in the browsing behavior of Web users with the objective to predict successive requests.

A further approach for sequential pattern discovery relies on the use of a tree structure. An application example of this kind of approach is represented by the Web Utilization Miner system [82]. In such system, transactions are derived from a collection of Web logs and transformed into sequences. Then sequences with the same prefix are merged into an aggregate tree. Each node in the tree structure represents a navigational subsequence from the root (an empty node) to a page and is annotated by the frequency of occurrences of that subsequence in the transaction data. Other examples of application of tree structures to discover sequential usage patterns have been described in [54, 69].

By using sequential pattern discovery, Web marketers can predict future visit patterns which will be helpful in placing advertisements aimed at certain user groups. Other types of temporal analysis that can be performed on sequential patterns include trend analysis, change point detection, or similarity analysis.

## Clustering

Clustering is the most widely employed technique in the pattern discovery process. Clustering is an unsupervised learning technique that tries to discover groups of similar items among large amount of data based on a distance function which computes the similarity between items. A comprehensive overview of works focusing on the use of Web data clustering methods to discover usage patterns is given in [87]. Two main types of interesting clusters can be discovered in the WUM domain, namely user clusters and Web document clusters.

The main aim of user clustering is to identify groups of users having similar browsing behavior. One of the first works where the authors suggested to shift the focus of WUM from single user to groups of user sessions is represented by [92]. Successively, a variety of clustering techniques have been employed in a large number of research works in order to discover significant usage patterns encoding the characteristics of the navigational behavior of user groups sharing the same interests. For example the well-known Fuzzy C-Means (FCM) algorithm has been successfully employed in [16] to discover similar behavior patterns from session data of users visiting a Web site. User sessions, expressed in the form of a matrix containing the interest degree of each identified user for each Web page, are partitioned by applying FCM into clusters representing user categories, i.e. groups of users with similar navigational behavior. FCM has been used in many other works to mine user categories from access data such as [6, 34].

A hierarchical clustering technique has been adopted in [50] to cluster users of a Web portal into a hierarchy of groups enclosing users characterized by a set of common interests. Then, for each group a prototype is determined that defines the profile of all group members.

In [9] a relational clustering approach has been applied. Based on this approach, relations among each pair of objects are collected into a relational dataset that is successively clustered for the identification of prototypes that can be interpreted as typical user interests. Other works have applied relational clustering algorithms for mining Web usage profiles. Among these, we mention the fuzzy c-Trimered Medoids Algorithm [34], the Fuzzy c-Medoids (FCMdd) algorithm [63], and the Relational Fuzzy Subtractive clustering algorithm [85].

Clustering of Web documents is devoted to discover groups of pages having related content. In [67] an exhaustive survey of Web document clustering approaches is given. Depending on the attributes of documents that are exploited in the clustering process, Web document clustering approaches can be categorized into three main classes, namely text-based, link-based, and hybrid.

In text-based approach, each Web document is characterized by its content, i.e. the words or the sentences contained in it. The key idea of such clustering approach is that two documents can be considered very similar if they have similar content (they contain many common words or sentences).

Apart from their content, Web documents are characterized by additional information such as metadata or hyperlinks that could be properly employed to partition them. In particular, link-based document clustering approaches employ information extracted by the link structure of pages to identify groups of similar pages. The basic idea of this approach is that if two documents are connected via a link, we may recognize a semantic relationship existing among them that could be exploited to partition documents.

Of course, the nature and the complexity of the link structure of Web pages highly affects the efficacy of link-based clustering algorithms. On the other hand, text-based algorithms result to have limits when dealing with some particularities of the language (such as synonyms, homonyms etc.). In addition, content of Web pages is not exclusively represented by text but it can be also expressed in other forms such as images or multimedia. As a consequence, hybrid document clustering approaches have been proposed in order to combine the advantages and to overcome the limitations of the previous approaches.


**Classification**

In contrast to clustering, classification is a supervised learning process that assigns an item into one of several predefined classes on the basis of certain features. In the specific context of WUM, classification has been used for modeling user behavior and the characteristics of Web pages in order to establish their membership to a particular category. In general, the classification process requires firstly the extraction and the selection of features able to describe the properties of the given classes. Then supervised learning algorithms are applied to classify items into predefined classes. This can be achieved by using learning techniques such as decision trees, naive Bayesian classifiers and neural networks.

A comparison among different learning algorithms based on these classification methods is described in [17].

A different classification approach is centered on the use of rough set theory. This approach involves the recognition of sets including similar items according to the attributes describing the available data. For example, rough set theory has been exploited to describe the user navigational behavior in [53]. In [45], a rough set based learning program has been applied to predict future user activities.

Unlike clustering, classification methods have found minor application in the WUM field. The limited use of classification is mainly due to the necessity of having preclassified data. This is not always possible when dealing with Web data, where items such as pages or sessions cannot always be labeled with categories. Thus unsupervised techniques, such as clustering, are revealed to be more suitable to be applied to a wider range of problems in the area of WUM.

## 4.2.4   Application of Web Usage Patterns

The overall goal of a generic WUM process is to mine usage patterns modeling the browsing behavior of Web users. The derived patterns can be properly analyzed in order to be exploited in a variety of application areas such as Web personalization, Web recommendation, Web site/page improvements, Web caching, etc. Of course the type of analysis of usage patterns is strongly dependent on the ultimate purposes of the specific application where patterns are exploited. In the following we discuss the most interesting areas where the employment of Web usage patterns turns to be useful to achieve the final goals of the considered Web application.

### Web Personalization

Web personalization is the process of customizing the content and/or the structure of a Web site to the needs/interests of users, taking advantage of the knowledge acquired from the analysis of their navigational behavior [56]. Personalization functions can be different according to the specific application in which these are integrated. An important example of personalization function consists in the capacity of a Web-based system to suggest interesting pages by visualizing a ranked list of links deemed useful for users in the current page. Complete surveys of works focusing on the use of usage-based approaches in the area of Web personalization are provided by [70], [56], [55]. Many research efforts have been addressed to the development of several personalization systems offering assistance and guidance functions to users visiting a site by exploiting usage information, as proved by many examples that can be retrieved in literature. Among these, WebWatcher [41] is a system that provides guidance to users by interactively suggesting hyperlinks. Such system is able to learn from user actions to acquire greater expertise for the topics chosen by the users in the past. A software agent named SiteHelper [66] acts as a helper for a Web user to find relevant information in a particular site. To assist the user in finding relevant information, the agent interactively and incrementally learns about the areas of interest of users and aids them accordingly. Letizia [51] is an agent that tracks the user browsing behavior (following links, initiating searches, requests for help) and tries to anticipate what items may be of interest for the same user.

Recent studies have investigated the possibility to include semantic concepts to enhance the usage mining process [84, 91]. For example in [68] the authors present SEWeP (SEmantic Web Personalization), a personalization system based on the combination of usage data and content semantics expressed in ontology terms, that is able to compute semantically enhanced navigational patterns and effectively generate useful suggestions. An overview of approaches embedding semantic knowledge into the usage-based Web personalization process is given in [26].

## Web Recommendation

Web Recommendation represents a promising technology that attempts to predict the interests of users, by providing them with products or services that they need without explicitly asking for them [59, 85]. Since recommendation systems enable the adaptation of Web sites to the characteristics of their users, they can be considered as a particular instance of personalization system. The development of recommender systems has become dramatically important in different application contexts, especially in e-commerce. In this specific context such systems are aimed at enhancing e-commerce sales by helping customers to find the desired products, suggesting additional products and improving customer loyalty. In [77] a survey of existing commercial recommendation systems implemented in e-commerce applications is presented. Among these, Amazon represents one of the most popular examples of e-commerce sites where several recommendation functions have been integrated in order to suggest additional products to a customer based on his/her previous purchases and browsing behavior. WebCF-DT [46] is a recommendation procedure that suggests products based on Web usage data as well as product purchase information and customer related data. E-bay, MovieFinder, MovieLens provide other important examples of e-commerce applications equipped with recommendation functionalities such as personalized shopping and suggestion of products satisfying customer needs.

## Web Site Improvements

The analysis of the Web usage patterns discovered by a WUM process may give important hints for enhancing the design of Web sites by (re)organizing their Web pages and improving the usability and accessibility of these. In this application context, many automatic ways have been developed with the aim to build sites able to reconfigure itself based on past behavior of the individual user or other users with similar profiles. This has led to the development of adaptive Web sites that are systems able to automatically improve their organization and presentation by learning from user access patterns. The study of the popularity of links in relation to their locations in the web page has become a fundamental requirement for different applications. The integration of this study in Web systems not only allows to obtain information about the most popular links of the site but also to suggest an optimal page design based on the frequency of the visits and the time spent on these links, hence adding a novel improvement to adaptive Web systems [17]. An exhaustive discussion on the literature works focusing on the use of WUM techniques to design Web adaptive systems is provided in [48].

**Web Caching**

Usage patterns obtained as a result of a WUM process may also provide useful insights into Web traffic behavior, allowing to develop adequate policies for Web Caching. Web Caching is the mechanism that consists in temporary storing Web pages requested by users for successive retrieval. Such mechanism offers different advantages such as the reduced latency and the reduced bandwidth consumption, the reduced server load [11]. In [49] an algorithm to make Web servers "pushier" is proposed. In this work, a set of association rules mined from access log data is used to determine the page that has to be prefetched. Thus, for example if the rule in the form "Page1 $\Rightarrow$ Page2" has been identified and selected, the Web server decides to prefetch "Page2" once "Page1" is requested. Many other methods for Web page prefetching based on access log information have been developed; examples are those proposed in [61, 78].

## 4.3   A Fuzzy Relational Clustering Technique for Mining User Categories

In this section, we describe a usage-based approach for the discovery of user categories describing the characteristics of groups of users having common interests. In our approach access log data are preprocessed and analyzed in order to mine a collection of significant patterns that model the user browsing behaviors. Hence, a fuzzy relational clustering technique is applied to the discovered patterns to derive a number of clusters representing user categories that encode interests shared by groups of users.

In this work, access log data are preprocessed by means of LODAP, a software tool that we have implemented for the analysis of Web log files in order to derive models characterizing the user browsing behaviors. To achieve this aim, LODAP executes the user session identification task devoted to the derivation of a set of user sessions by determining the sequence of pages accessed by each user during a predefined time period. Then user sessions are exploited to create models expressing the interest degree exhibited by each user for each visited page of the site. Briefly speaking, LODAP performs log file preprocessing through four main steps:

1. *Data Cleaning* that removes all redundant and useless records contained in the Web log file (e.g. accesses to multimedia objects, robots' requests, etc.) so as to retain only information concerning accesses to pages of the Web site.
2. *Data Structuration* that groups the significant requests into user sessions. Each user session contains the sequence of pages accessed by the same user during an established time period.
3. *Data Filtering* that selects only significant pages accessed in the Web site. In this step, the least visited pages as well as the most visited ones, are removed.

4. *Interest degree computation* that exploits information about accessed pages to create a model of the user behavior by evaluating a degree of interest of each user for each accessed page.

Interested readers can find major details about LODAP in [14].

User behavior models extracted by LODAP are synthetized in a behavior matrix $\mathbf{B} = [b_{ij}]$ where the rows $i = 1, \ldots, n$ represent the users and the columns $j = 1, \ldots, m$ correspond to the Web pages of the site. Each component $b_{ij}$ of the matrix indicates the interest degree of the $i$-th user for the $j$-th page. The $i$-th user behavior vector $\mathbf{b}_i$ ($i$-th row of the behavior matrix) characterizes the browsing behavior of the $i$-th user.

Successively, starting from the user behavior data, we apply CARD+, a fuzzy relational clustering algorithm in order to categorize users. In the categorization process, two main activities may be distinguished:

- The creation of the relation matrix by computing the dissimilarity values among all pairs of users;
- The categorization of users by grouping similar users into categories.

In the following subsections we detail the activities performed in the categorization process of Web users.


### 4.3.1 Computing (dis)similarity among Web Users

The first activity in the user categorization process based on the use of a relational fuzzy clustering technique consists in the creation of the relation matrix including the dissimilarity values among all pairs of users. To create the relation matrix, an essential task is the evaluation of the (dis)similarities among two generic users on the basis of a proper measure. In our case, based on the behavior matrix, the similarity between two generic users is expressed by the similarity between the two corresponding user behavior vectors.

In literature, different metrics have been proposed to measure the similarity degree between two generic objects. One of the most common measures employed to this aim is the angle cosine measure [73]. In the specific context of user category extraction, the similarity among any two behavior vectors $\mathbf{b}_x$ and $\mathbf{b}_y$ expressed by the cosine measure is defined as follows:

$$Sim_{Cos}\left(\mathbf{b}_x, \mathbf{b}_y\right) = \frac{\mathbf{b}_x \mathbf{b}_y'}{\|\mathbf{b}_x\| \|\mathbf{b}_y\|} = \frac{\sum_{j=1}^m b_{xj} b_{yj}}{\sqrt{\sum_{j=1}^m b_{xj}^2} \sqrt{\sum_{j=1}^m b_{yj}^2}}. \tag{4.1}$$

The use of the cosine measure to define the similarity between two users visiting a Web site could produce ineffective results since it takes into account only the common pages visited by the considered users. In this way, we might have the loss of semantic information underlying Web usage data related to the relevance of each page for each user. Thus, to better capture the similarity between two generic Web users, we defined a fuzzy similarity measure. Specifically, two generic users are modeled as two fuzzy sets and the similarity between these users is defined as the similarity between the corresponding fuzzy sets. To do so, the user behavior matrix **B** is converted into a matrix $\mathbf{M} = [\mu_{ij}]$ which expresses the interest degree of each user for each page in a fuzzy way. A simple characterization of the matrix **M** is provided as follows:

$$\mu_{ij} = \begin{cases} 0 & \text{if} \quad b_{ij} < ID_{min} \\ \frac{b_{ij}-ID_{min}}{id_{max}-ID_{min}} & \text{if } b_{ij} \in [ID_{min}, ID_{max}] \\ 1 & \text{if} \quad b_{ij} > ID_{max} \end{cases} \quad (4.2)$$

where $ID_{min}$ is a minimum threshold for the interest degree under which the interest for a page is considered null, and $ID_{max}$ is a maximum threshold of the interest degree above which the page is considered surely preferred by the user.

Starting from this fuzzy characterization, the rows of the new matrix **M** are interpreted as fuzzy sets defined on the set of Web pages. Each fuzzy set $\mu_i$ is related to a user $\mathbf{b}_i$ and it is simply characterized by the following membership function:

$$\mu_i(j) = \mu_{ij} \quad \forall j = 1, 2, \ldots, m \quad (4.3)$$

In this way the similarity of two generic users is intuitively defined as the similarity between the corresponding fuzzy sets. The similarity among fuzzy sets can be evaluated in different ways [89]. One of the most common measures to evaluate similarity between two fuzzy sets is the following:

$$\sigma(\mu_1, \mu_2) = \frac{|\mu_1 \cap \mu_2|}{|\mu_1 \cup \mu_2|} \quad (4.4)$$

According to this measure, the similarity between two fuzzy sets is given by the ratio of two quantities, namely the cardinality of the intersection of the fuzzy sets and the cardinality of the union of the fuzzy sets. The intersection of two fuzzy sets is defined by the minimum operator, as follows:

$$(\mu_1 \cap \mu_2)(j) = \min\{\mu_{\mathbf{b}_1}(j)\mu_{\mathbf{b}_2}(j)\} \quad (4.5)$$

The union of two fuzzy sets is defined by the maximum operator, as follows:

$$(\mu_1 \cup \mu_2)(j) = \max\{\mu_{\mathbf{b}_1}(j)\mu_{\mathbf{b}_2}(j)\} \quad (4.6)$$

The cardinality of a fuzzy set (also called "$\sigma$-count") is computed by summing up all its membership values:

$$|\mu| = \sum_{j=1}^{m} \mu(j) \quad (4.7)$$

Summarizing, the similarity between any two users $\mathbf{b}_x$ and $\mathbf{b}_y$ is defined as follows:

$$Sim_{fuzzy}\left(\mathbf{b}_x, \mathbf{b}_y\right) = \frac{\sum_{j=1}^{m} \min\left\{\mu_{\mathbf{b}_x j}, \mu_{\mathbf{b}_y j}\right\}}{\sum_{j=1}^{m} \max\left\{\mu_{\mathbf{b}_x j}, \mu_{\mathbf{b}_y j}\right\}}. \tag{4.8}$$

This fuzzy similarity measure allows to embed the semantic information incorporated in the user behavior data. In this way an actual estimation of the similarity degree between two user behaviors is obtained. Similarity values are mapped into the similarity matrix $\mathbf{Sim} = \left[Sim_{ij}\right]_{i,j=1,\ldots,n}$ where each component $Sim_{ij}$ expresses the similarity value between the user behavior vectors $\mathbf{b}_i$ and $\mathbf{b}_j$ calculated by using the fuzzy similarity measure. Starting from the similarity matrix, the dissimilarity values are simply computed as $Diss_{ij} = 1 - Sim_{ij}$, for $i, j = 1, \ldots, n$. These are mapped in a $n \times n$ matrix $\mathbf{R} = \left[Diss_{ij}\right]_{i,j=1,\ldots,n}$ representing the relation matrix.

### 4.3.2   Grouping Users by Fuzzy Clustering

The next activity of the categorization process is the clustering of user behaviors in order to group users with similar interests into a number of user categories. To accomplish this, we employ CARD+ that we proposed in [15] as an improved version of the CARD (Competitive Agglomeration Relational Data) clustering algorithm [62]. The main feature of CARD+ is its ability to automatically categorize the available data into an optimal number of clusters starting from an initial random number. In our experience CARD resulted to be very sensitive to the initial number of clusters by often providing different final partitions, thus failing in finding the actual number of clusters buried in data. Indeed we observed that CARD produces redundant partitions, with clusters having a high overlapping degree (very low inter-cluster distance). CARD+ overcomes this limitation by adding a post-clustering process to the CARD algorithm in order to remove redundant clusters.

According to the relational clustering approach, CARD+ obtains an implicit partition of the object data by deriving the distances from the relational data to a set of $C$ implicit prototypes that summarize the data objects belonging to each cluster in the partition. Specifically, starting from the relation matrix $\mathbf{R}$, the following implicit distances are computed at each iteration step of the algorithm:

$$d_{ci} = (\mathbf{R}\mathbf{z}_c)_i - \mathbf{z}_c\mathbf{R}\mathbf{z}_c/2 \tag{4.9}$$

for all behavior vectors $i = 1, \ldots, n$ and for all clusters $c = 1, \ldots, C$, where $\mathbf{z}_c$ is the membership vector for the $c$-th cluster defined on the basis of the fuzzy membership values $z_{ci}$ that describe the degree of belongingness of the $i$-th behavior vector in the $c$-th cluster. Once the implicit distance values $d_{ci}$ have been computed, the fuzzy membership values $z_{ci}$ are updated to optimize the clustering criterion, resulting in a new fuzzy partition of behavior vectors. The process is iterated until the membership values stabilize.

Finally, a crisp assignment of behavior vectors to the identified clusters is performed in order to derive a prototype vector for each cluster, representing a user category. Precisely, each behavior vector is crisply assigned to the closest cluster, creating $C$ clusters:

$$\chi_c = \{\mathbf{b}_i \in \mathbf{B} | d_{ci} < d_{ki} \forall c \neq k\} \quad 1 \leq c \leq C. \tag{4.10}$$

Then, for each cluster $\chi_c$ a prototype vector $\mathbf{v}_c = (v_{c1}, v_{c2}, \ldots, v_{cm})$ is derived, where

$$v_{cj} = \frac{\sum_{\mathbf{b}_i \in \chi_c} b_{ij}}{|\chi_c|} \quad j = 1, \ldots, N_P. \tag{4.11}$$

The values $v_{cj}$ represent the significance (in terms of relevance degree) of a given page $p_j$ to the $c$-th user category.

Summarizing, the CARD+ mines a collection of $C$ clusters from behavior data, representing categories of users that have accessed to the Web site under analysis. Each category prototype $\mathbf{v}_c = (v_{c1}, v_{c2}, \ldots, v_{cm})$ describes the typical browsing behavior of a group of users with similar interests about the most visited pages of the Web site.

## 4.3.3   A Case Study

To show the suitability of CARD+ equipped with the fuzzy measure to identify Web user categories, we carried out an experimental simulation. We used the access logs from a Web site targeted to young users (average age 12 years old), i.e. the Italian Web site of the Japanese movie Dragon Ball (www.dragonballgt.it). This site was chosen because of its high daily number of accesses (thousands of visits each day).

### Preprocessing Log Files

Firstly the preprocessing of log files was executed to derive models of user behavior. LODAP was used to identify user behavior vectors from the log data collected during a period of 12 hours (from 10:00 a.m. to 22:00 p.m.). Once the four steps of LODAP were executed, a 200 × 42 behavior matrix was derived. The 42 pages in the Web site were labeled with a number (see table 1) to facilitate the analysis of results, by specifying the content of the Web pages.

### Grouping Web Users

Starting from the available behavior matrix, the relation matrix was created by using the fuzzy similarity measure.

**Table 4.1** Description of the retained pages in the Web site

| Pages | Content |
| --- | --- |
| 1, ..., 8 | Pictures of characters |
| 9,..., 13 | Various kind of pictures related to the movie |
| 14,..., 18 | General information about the main character |
| 19, 26, 27 | Matches |
| 20, 21, 36 | Services (registration, login, ...) |
| 22, 23, 24, 25, 28, ..., 31 | General information about the movie |
| 32, ..., 37 | Entertainment (games, videos,...) |
| 38, ..., 42 | Description of characters |

Next, the CARD+ algorithm (implemented in the Matlab environment 6.5) was applied to the behavior matrix in order to obtain clusters of users with similar browsing behavior. We carried out several runs by setting a different initial number of clusters $C_{max}$ = (5, 10, 15). To establish the goodness of the derived partitions of behavior vectors, at the end of each run two indexes were calculated: the Dunn's index and the Davies-Bouldin index [35]. These were used in different works to evaluate the compactness of the partitions obtained by several clustering algorithms. Good partitions correspond to large values of the Dunn's index and low values for the Davies-Bouldin index. We observed that CARD+ with the use of the fuzzy similarity measure provided data partitions with the same final number of clusters $C = 5$, independently from the initial number of clusters $C_{max}$. The validity indexes took the same values in all runs. In particular, the Dunn's index value was always equal to 1.35 and the value for the Davies-Bouldin index was 0.13. As a consequence the CARD+ algorithm equipped with the fuzzy similarity measure resulted to be quite stable, by partitioning the available behavior data into 5 clusters corresponding to the identified user categories.

**Experimental Results**

To evaluate the effectiveness of the employed fuzzy similarity measure, we compared it to the cosine measure within the CARD+ algorithm. We carried out the same trials of the previous experiments. Moreover to establish the suitability of CARD+ for the task of user categorization, we applied the original CARD algorithm to categorize user behaviors by employing either the cosine measure and the fuzzy similarity measure for the computation of the relation matrix. In figures 4.1 and 4.2 the obtained values for the validity indexes are compared. In these figures the final number of clusters extracted by the employed clustering algorithm is also indicated. As it can be observed, CARD+ with the use of the cosine measure derived 4 or 5 clusters, resulting less stable than CARD+ equipped with the fuzzy similarity measure. Moreover the CARD algorithm showed an instable behavior with both the

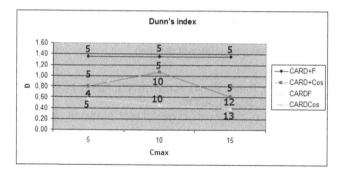

**Fig. 4.1** Comparison of the Dunn's index obtained by the employed algorithms and similarity measures

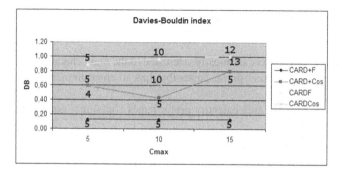

**Fig. 4.2** Comparison of the Davies-Bouldin index obtained by the employed algorithms and similarity measures

similarity measures, by providing data partitions with a different final number of clusters in each trial.

Analyzing the results obtained by the different runs, we can conclude that CARD+ with the employment of the fuzzy similarity measure was able to derive the best partition in terms of compactness; hence, it revealed to be a valid approach for the identification of user categories.

The information about the user categories extracted by CARD+ equipped with the fuzzy similarity measure are summarized in table 2. In particular, for each user category (labeled with numbers 1,2,...,5) the pages with the highest degree of interest are indicated. It can be noted that some pages (e.g. $P_1$, $P_2$, $P_3$, $P_{10}$, $P_{11}$, and $P_{12}$) are included in more than one user category, showing how different categories of users may exhibit common interests.

We can give an interpretation of the identified user categories, as follows:

- Category 1. Users in this category are mainly interested on information about the movie characters.
- Category 2. Users in this category are interested in the history of the movie and in pictures of movie and characters.

**Table 4.2** User categories identified by CARD+ equipped with the fuzzy similarity measure

| User category | Relevant pages (interest degrees) |
|---|---|
| 1 | $P_1(55)$, $P_2(63)$, $P_3(54)$, $P_5(52)$, $P_7(48)$, $P_8(43)$, $P_{14}(66)$, $P_{28}(56)$, $P_{29}(52)$, $P_{30}(37)$ |
| 2 | $P_1(72)$, $P_2(59)$, $P_3(95)$, $P_6(65)$, $P_7(57)$, $P_{10}(74)$, $P_{11}(66)$, $P_{13}(66)$ |
| 3 | $P_1(50)$, $P_2(50)$, $P_3(45)$, $P_4(46)$, $P_5(42)$, $P_6(42)$, $P_8(34)$, $P_9(37)$, $P_{12}(40)$, $P_{15}(41)$, $P_{16}(41)$, $P_{17}(38)$, $P_{18}(37)$, $P_{19}(36)$ |
| 4 | $P_2(49)$, $P_{10}(47)$, $P_{11}(38)$, $P_{12}(36)$, $P_{14}(27)$, $P_{31}(36)$, $P_{32}(29)$, $P_{33}(39)$, $P_{34}(36)$, $P_{35}(26)$, $P_{36}(20)$, $P_{37}(37)$, $P_{38}(29)$, $P_{39}(30)$, $P_{40}(34)$, $P_{41}(28)$, $P_{42}(24)$ |
| 5 | $P_4(70)$, $P_5(65)$, $P_{20}(64)$, $P_{21}(62)$, $P_{22}(54)$, $P_{23}(63)$, $P_{24}(54)$, $P_{25}(41)$, $P_{26}(47)$, $P_{27}(47)$ |

- Category 3. These users are mostly interested to the main character of the movie.
- Category 4. These users prefer pages that link to entertainment objects (games and video).
- Category 5. Users in this category prefer pages containing general information about the movie.

The extracted user categories may be used to implement personalization functions in the considered Web site such as, for example, the suggestion of interesting links to users visiting the site, as described in the next section.

### Exploiting User Categories

Once user categories have been discovered, these can be exploited to embed personalization functionalities into the considered Web site offering useful assistance to users. An example of personalization function is the dynamic suggestion of links deemed interesting to current users taking into account their preferences and needs. This function may be realized in a straightforward manner by visualizing in a portion of the currently visited page a ranked list of links retained useful according to the membership values of the user to the identified categories.

In particular, the semantic interpretation associated to each extracted user category makes possible the identification of a set of pages (corresponding to those pages with highest relevance degrees in each category prototype) that can be considered relevant for users belonging to the corresponding category. Whenever a new user visits the site, an on-line module matches his/her current partial session against the set of user category prototypes currently available to derive his/her membership values to the different user categories. Based on these values, we may generate a recommendation list which will include a set of ordered links reflecting the actual interests of the current user.

To determine the list of links to be recommended, the *K Nearest Neighbors* approach followed by top-$N$ recommendations can be adopted. Based on this approach, given a current session, the closest $K$ user categories are found according to certain measures able to evaluate the similarity between current session of the user and user category prototypes. Then, links associated to the $K$ nearest user categories are sorted in decreasing order of their relevance, and the top-$N$ links are treated as recommendation list that can be presented in the page visualized by the current user.

## 4.4  Conclusions

With the growth of Web-based applications, significant interest has been addressed to the discovery and the analysis of Web usage patterns. Extracting models characterizing the browsing behavior of users and applying the discovered knowledge may help users during their interactions with the Web. In this work we mainly focused on the use of WUM approaches for mining usage patterns. In particular we provided a comprehensive overview of a general WUM process highlighting the main issues related to each stage involved in the process, along with the most commonly employed techniques. Moreover we discussed the main areas that can benefit from the use of Web usage patterns. As an illustrative example, we described a WUM approach that exploits a fuzzy relational clustering technique embedding a fuzzy measure to evaluate the similarity among users for the discovery of user categories reflecting the common characteristics of groups of users showing the same interests when visiting a Web site. The presented WUM approach is used to stress the effectiveness of clustering techniques in identification of significant patterns encoding the actual interests of users. Also, it highlights the need to define new and appropriate measures for computing the similarity among users.

WUM has become a well established field of research, as confirmed by the large number of works published on this topic and the variety of contexts where WUM processes can be conveniently applied. However much work can be still done to enhance the efficacy of the overall process of discovery and analysis of usage patterns. One of the most interesting and recent research directions concerns the possibility to integrate semantics within Web site design with the aim to improve the results of WUM applications. Efforts in this direction seem to be a fruitful way to create much more effective WUM based systems that are consistent with the emergence and proliferation of the Semantic Web.

## References

1. Abraham, A.: Business intelligence from web usage mining. Journal of Information & Knowledge Management 2(4), 375–390 (2003)
2. Abraham, A.: i-Miner: A web usage mining framework using hierarchical intelligent systems. In: Proc. of the IEEE International Conference on Fuzzy Systems (FUZZ-IEEE 2003), pp. 1129–1134 (2003)

3. Agrawal, R., Imielinski, T., Swami, A.N.: Mining association rules between sets of items in large databases. In: Proc. of the 1993 ACM SIGMOD International Conference on Management of Data (SIGMOD 1993), pp. 207–216 (1993)
4. Agrawal, R., Srikant, R.: Mining sequential patterns. In: Proc. of the Eleventh International Conference on Data Engineering (ICDE 1995), pp. 3–14 (1995)
5. Anderson, C.R., Domingos, P., Weld, D.S.: Adaptive Web Navigation for Wireless Devices. In: Proc. of the 17th International Joint Conference on Artificial Intelligence (IJCAI 2001), pp. 879–884 (2001)
6. Arotariteia, D., Mitra, S.: Web mining: a survey in the fuzzy framework. Fuzzy Sets and Systems 148(1), 5–19 (2004)
7. Bayir, M.A., Cosar, A., Toroslu, I.H., Fidan, G.: Smart Miner: A New Framework for Mining Large Scale Web Usage Data. In: Proc. of the 18th International Conference on World Wide Web, pp. 161–170 (2009)
8. Berendt, B.: Web usage mining, site semantics, and the support of navigation. In: Proc. of Workshop Web Mining for E-Commerce - Challenges and Opportunities, pp. 83–93 (2000)
9. Runkler, T.A., Bezdek, J.C.: Web mining with relational clustering. International Journal of Approximate Reasoning 32, 217–236 (2003)
10. Borges, J.A., Levene, M.: Generating Dynamic Higher-Order Markov Models in Web Usage Mining. In: Jorge, A.M., Torgo, L., Brazdil, P.B., Camacho, R., Gama, J. (eds.) PKDD 2005. LNCS (LNAI), vol. 3721, pp. 34–45. Springer, Heidelberg (2005)
11. Davison, B.D.: A Web caching primer. IEEE Internet Computing 5(4), 38–45 (2001)
12. Buchner, A.G., Mulvenna, M.D.: Discovering internet marketing intelligence through online analytical web usage mining. SIGMOD Record 27(4), 54–61 (1999)
13. Cadez, I., Heckerman, D., Meek, C., Smyth, P., White, S.: Visualization of Navigation Patterns on a Web Site Using Model Based Clustering. Technical Report MSR-TR-00-18 (2000)
14. Castellano, G., Fanelli, A.M., Mencar, C., Torsello, M.A.: Log data preprocessing for mining Web browsing patterns. In: Proc. of the 8th Asian Pacific Industrial Engineering and Management Systems Conference (APIEMS 2007) (2007)
15. Castellano, G., Fanelli, A.M., Torsello, M.A.: Relational Fuzzy approach for Mining User Profiles. In: Aggarwal, A., Yager, R., Sandeberg, I.W. (eds.) Lectures Notes in Computational Intelligence, pp. 175–179. Wseas Press (2007)
16. Castellano, G., Mesto, F., Minunno, M., Torsello, M.A.: Web User Profiling Using Fuzzy Clustering. In: Masulli, F., Mitra, S., Pasi, G. (eds.) WILF 2007. LNCS (LNAI), vol. 4578, pp. 94–101. Springer, Heidelberg (2007)
17. Chan, P.K.: A non-invasive learning approach to building Web user profiles. In: Proc. of 5th ACM SIGKDD International Conference, Workshop on Web Usage Analysis and User Profiling, pp. 7–12 (1999)
18. Chen, P., Kuo, F.: An information retrieval system based on an user profile. The Journal of Systems and Software 54, 3–8 (2000)
19. Chitraa, V., Davamani, A.S.: A Survey on Preprocessing Methods for Web Usage Data. International Journal of Computer Science and Information Security 7(3), 78–83 (2010)
20. Cho, Y., Kim, J.K., Kim, S.H.: A personalized recommender system based on web usage mining and decision tree induction. Expert Systems with Applications 23(3), 329–342 (2003)
21. Cimiano, P., Staab, S.: Learning by googling. SIGKDD Explorations Newsletter 6(2), 24–33 (2004)

22. Cohen, E., Krishnamurthy, B., Rexford, J.: Improving end-to-end performance of the web using server volumes and proxy filters. In: Proc. of Conference on Applications, Technologies, Architectures, and Protocols for Computer Communication (ACM SIG-COMM 1998), pp. 241–253 (1998)

23. Cooley, R., Mobasher, B., Srivastava, J.: Data preparation for mining world wide web browsing patterns. Knowledge and Information Systems. 1(1), 55–32 (1999)

24. Cooley, R.: Web usage mining: discovery and application of interesting patterns from Web data. PhD thesis, University of Minnesota (2000)

25. Costa, M., Gong, Z.: Web structure mining: an introduction. In: Proc. of the IEEE International Conference on Information Acquisition, pp. 590–595 (2005)

26. Dai, H., Mobasher, B.: Using ontologies to discover domain-level web usage profiles. In: Proc. of the 2nd Semantic Web Mining Workshop (2002)

27. Facca, F.M., Lanzi, P.: Mining interesting knowledge from weblogs: a survey. Data & Knowledge Engineering 53, 225–241 (2005)

28. Frías-Martínez, E., Karamcheti, V.: A Customizable Behavior Model for Temporal Prediction of Web User Sequences. In: Zaïane, O.R., Srivastava, J., Spiliopoulou, M., Masand, B. (eds.) WebKDD 2003. LNCS (LNAI), vol. 2703, pp. 66–85. Springer, Heidelberg (2003)

29. Furnkranz, J.: Web structure mining - exploiting the graph structure of the world-wide web. GAI-Journal 21(2), 17–26 (2002)

30. Furnkranz, J.: Web mining. In: Maimon, O., Rokach, L. (eds.) Data Mining and Knowledge Discovery Handbook, pp. 899–920. Springer (2005)

31. Ghorbani, A.A., Xu, X.: A fuzzy markov model approach for predicting user navigation. Web Intelligence, 307–311 (2007)

32. Godoy, D., Amandi, A.: Learning browsing patterns for context-aware recommendation. In: Bramer, M. (ed.) Artificial Intelligence in Theory and Pratice, pp. 61–70. Springer (2006)

33. Han, J., Kamber, M.: Data Mining Concepts and Techniques. Morgan Kaufmann (2001)

34. Krishnapuram, R., Joshi, A., Nasraoui, O., Yi, L.: Low-complexity fuzzy relational clustering algorithms for web mining. Journal IEEE-FS 9, 595–607 (2001)

35. Halkidi, M., Batistakis, Y., Vazirgiannis, M.: Cluster Validity Methods: Part II. SIGMOD Record 31(3), 19–27 (2002)

36. Hansen, M., Shriver, E.: Using navigation data to improve IR functions in the context of web search. In: Proc. of the 10th International Conference on Information and Knowledge Management, pp.135–142 (2001)

37. Heer, J., Chi, E.H.: Identification of web user traffic composition using multi-modal clustering and information scent. In: Proc. of the Workshop on Web Mining, SIAM Conference on Data Mining, pp. 51–58 (2001)

38. Huang, X., Cercone, N., An, A.: Comparison of interestingness functions for learning web usage patterns. In: Proc. of the 11th International Conference on Information and Knowledge Management, pp. 617–620 (2002)

39. Hussain, T., Asghar, S., Masood, N.: Web usage mining: A survey on preprocessing of web log file. In: Proc. of the International Conference on Information and Emerging Technologies (ICIET), pp. 1–6 (2010)

40. Jespersen, S.E., Thorhauge, J., Bach Pedersen, T.: A Hybrid Approach to Web Usage Mining. In: Kambayashi, Y., Winiwarter, W., Arikawa, M. (eds.) DaWaK 2002. LNCS, vol. 2454, pp. 73–82. Springer, Heidelberg (2002)

41. Joachims, T., Freitag, D., Mitchell, T.: Webwatcher: A tour guide for the world wide web. In: Proc. of the 15th International Conference on Artificial Intelligence, pp. 770–775 (1997)

42. Joshi, K., Joshi, A., Yesha, Y.: On using a warehouse to analyse web logs. Distributed and Parallel Databases 13(2), 161–180 (2003)
43. Kamdar, T., Joshi, A.: On creating adaptive web sites using web log mining. Technical report tr-cs-00-05. Department of Computer Science and Electrical Engineering University of Maryland (2000)
44. Khasawneh, N., Chan, C.-C.: Active user-based and ontology-based web log data preprocessing for web usage mining. In: Proc. of the 2006 IEEE/WIC/ACM International Conference on Web Intelligence, pp. 325–328 (2006)
45. Khasawneh, N., Chan, C.-C.: Web Usage Mining Using Rough Sets. In: Proc. of the 2005 Annual Meeting of the North American Fuzzy Information Processing Society, pp. 580–585 (2005)
46. Kima, J.K., Chob, Y.H., Kimc, W.J., Kimc, J.R., Suha, J.H.: A personalized recommendation procedure for Internet shopping support. Electronic Commerce Research and Applications 1, 301–313 (2002)
47. Kosala, R., Blockeel, H.: Web mining research: a survey. ACM SIGKDD Explorations Newsletter 2, 1–15 (2000)
48. Koutri, M., Avouris, N., Daskalaki, S.: A Survey of Web-Usage Mining: Techniques for Building Web-Based Adaptive Hypermedia Systems. In: Chen, S.Y., Magoulas, G.D. (eds.) Adaptable and Adaptive Hypermedia Systems, pp. 125–150. IRM Press (2005)
49. Lan, B., Bressan, S., Ooi, B.-C., Tay, Y.C.: Making Web Servers Pushier. In: Masand, B., Spiliopoulou, M. (eds.) WebKDD 1999. LNCS (LNAI), vol. 1836, pp. 112–125. Springer, Heidelberg (2000)
50. Lazzerini, B., Marcelloni, F.: A hierarchical fuzzy clustering-based system to create user profiles. International Journal on Soft Computing 11, 157–168 (2007)
51. Lieberman, H.: Letizia: An agent that assists web browsing. In: Proc. of the 1995 International Joint Conference on Artificial Intelligence, pp. 924–929 (1995)
52. Liu, B., Chang, K.C.C.: Editorial: Special issue on web content mining. SIGKDD Explorations special issue on Web Content Mining 6(2), 1–4 (2004)
53. Maheswari, V.U., Siromoney, A., Mehata, K.M.: The Variable Precision Rough Set Model for Web Usage Mining. In: Zhong, N., Yao, Y., Ohsuga, S., Liu, J. (eds.) WI 2001. LNCS (LNAI), vol. 2198, pp. 520–524. Springer, Heidelberg (2001)
54. Menasalvas, E., Millan, S., Pena, J., Hadjimichael, M., Marban, O.: Subsessions: a granular approach to click path analysis. In: Proc. of the FUZZ-IEEE Fuzzy Sets and Systems Conference, pp. 12–17 (2002)
55. Mobasher, B.: Web usage mining and personalization. In: Singh, M.P. (ed.) Practical Handbook of Internet Computing, pp. 1–35. CRC Press (2005)
56. Mobasher, B., Cooley, R., Srivastava, J.: Automatic personalization based on web usage mining. Communications of the ACM 43(8), 142–151 (2000)
57. Mobasher, B.: Web usage mining. In: Liu, B. (ed.) Web Data Mining: Exploring Hyperlinks, Contents and Usage Data, pp. 449–483. Springer, Heidelberg (2006)
58. Mortazavi-Asl, B.: Discovering and mining user web-page traversal patterns. Masters thesis, Simon Fraser University (2001)
59. Mulvenna, M., Anand, S., Buchner, A.: Personalization on the net using Web mining CACM, vol. 43, pp. 123–125 (2000)
60. Nanopoulos, A., Katsaros, D., Manolopoulos, Y.: Exploiting Web Log Mining for Web Cache Enhancement. In: Kohavi, R., Masand, B., Spiliopoulou, M., Srivastava, J. (eds.) WebKDD 2001. LNCS (LNAI), vol. 2356, pp. 68–87. Springer, Heidelberg (2002)
61. Nanopoulos, A., Katsaros, D., Manolopoulos, Y.: Effective prediction of Web-user accesses: A data mining approach. In: Proc. of the 3rd International Workshop on Mining Web Log Data Across (2001)

62. Nasraoui, O., Frigui, H., Joshi, A., Krishnapuram, R.: Mining Web access log using relational competitive fuzzy clustering. Journal of Computer Engineering 1, 195–204 (1999)

63. Nasraoui, O., Krishnapuram, R., Joshi, A., Kamdar, T.: Automatic Web User Profiling and Personalization using a Robust Fuzzy Relational Clustering. In: Segovia, J., Szczepaniak, P., Niedzwiedzinski, M. (eds.) E-Commerce and Intelligent Methods in Studies in Fuzziness and Soft Computing. Springer (2002)

64. Nasraoui, O., Petenes, C.: Combining web usage mining and fuzzy inference for website personalization. In: Proc. of Workshop on Web Mining and Web Usage Analysis, pp. 37–46 (2003)

65. Nasraoui, O., Soliman, M., Saka, E., Badia, A., Germain, R.: A web usage mining framework for mining evolving user profiles in dynamic web sites. IEEE Transaction on Knowledge Data Engineering 20(2), 202–215 (2008)

66. Ngu, D.S.W., Wu, X.: Sitehelper: A localized agent that helps incremental exploration of the world wide web. In: Proc. of the 6th International World Wide Web Conference, pp. 1249–1255 (1997)

67. Oikonomakou, N., Vazirgiannis, M.: A Review of Web Document Clustering Approaches. In: Maimon, O., Rokach, L. (eds.) Data Mining and Knowledge Discovery Handbook, pp. 921–943. Springer (2005)

68. Paulakis, S., Lampos, C., Eirinaki, M., Vazirgiannis, M.: SEWeP: A Web Mining System Supporting Semantic Personalization. In: Boulicaut, J.-F., Esposito, F., Giannotti, F., Pedreschi, D. (eds.) PKDD 2004. LNCS (LNAI), vol. 3202, pp. 552–554. Springer, Heidelberg (2004)

69. Pei, J., Han, J., Motazavi-Asl, B., Zhu, H.: Mining access patterns efficiently from web logs. In: Proc. of the Pacific-Asia Conference on Knowledge Discovery and Data Mining, pp. 396–407 (2000)

70. Pierrakos, D., Paliouras, G., Papatheodorou, C., Spyropoulos, C.D.: Web usage mining as a tool for personalization: a survey. User Modeling and User-Adapted Interaction 13(4), 311–372 (2003)

71. Piramuthu, S.: On learning to predict web traffic. Decision Support Systems 35(2), 213–229 (2003)

72. Pitkow, J.: In search of reliable usage data on the WWW. In: Proc. of the 6th Int. World Wide Web Conference, pp. 451–463 (1997)

73. Rossi, F., De Carvalho, F., Lechevallier, Y., Da Silva, A.: Dissimilarities for Web Usage Mining. In: Batagelj, V., Hans-Hermann, B., Ferligoj, A., Ziberna, A. (eds.) Data Science and Classification, Studies in Classification, Data Analysis and Knowledge Organization, pp. 39–46. Springer (2006)

74. Roussinov, D., Zhao, J.L.: Automatic discovery of similarity relationships through web mining. Decision Support Systems 35(1), 149–166 (2003)

75. Sarwar, B.M., Karypis, G., Konstan, J., Riedl, J.: Analysis of recommender algorithms for e-commerce. In: Proc. of the 2nd ACM E-Commerce Conference (EC 2000), pp. 158–167 (2000)

76. Sathiyamoorthi, V., Murali Bhaskaran, V.: Data Preparation Techniques for Web Usage Mining in World Wide Web-An Approach. International Journal of Recent Trends in Engineering 2(4), 1–4 (2009)

77. Schafer, J.B., Konstan, J.A., Riedl, J.: E-commerce recommendation applications. Data Mining and Knowledge Discovery 5(1-2), 115–153 (2001)

78. Schechter, S.E., Krishnan, M., Smith, M.D.: Using path profiles to predict HTTP requests. In: Proc. of the 7th International World Wide Web Conference, pp. 457–467 (1998)

104 G. Castellano, A.M. Fanelli, and M.A. Torsello

79. Shahabi, C., Banaei-Kashani, F., Faruque, J.: A reliable, efficient, and scalable system for web usage data acquisition. In: Proc. of WEBKDD 2001 Mining Log Data Across All Customer Touch Points (2001)
80. Spilipoulou, M., Mobasher, B., Berendt, B.: A framework for the Evaluation of Session Reconstruction Heuristics in Web Usage Analysis. INFORMS Journal on Computing Spring 15(2), 171–190 (2003)
81. Spiliopoulou, M.: Data mining for the web. In: Proc. of the 3rd European Conference on Principles and Practice of Knowledge Discovery in Databases, pp. 588–589 (1999)
82. Spiliopoulou, M., Faulstich, L.C.: WUM: A Web Utilization Miner. In: Proc. of the International Workshop on the Web and Databases, pp. 109–115 (1999)
83. Srivastava, J., Cooley, R., Deshpande, M., Tan, P.-N.: Web usage mining: Discovery and applications of usage patterns from web data. SIGKDD Explorations 1(2), 1–12 (2000)
84. Stumme, G., Hotho, A., Berendt, B.: Usage Mining for and on the Semantic Web. Methods, pp. 461–481. AAAI Press (2004)
85. Suryavanshi, B., Shiri, N., Mudur, S.: An efficient technique for mining usage profiles using relational fuzzy subtractive clustering. In: Proc. of the 2005 International Workshop on Challenges in Web Information Retrieval and Integration (WIRI 2005), pp. 23–29 (2005)
86. Tan, P.N., Kumar, V.: Discovery of web robot sessions based on their navigational patterns. Data Mining and Knowledge Discovery 6(1), 9–35 (2002)
87. Vakali, A.I., Pokorný, J., Dalamagas, T.: An Overview of Web Data Clustering Practices. In: Lindner, W., Fischer, F., Türker, C., Tzitzikas, Y., Vakali, A.I. (eds.) EDBT 2004. LNCS, vol. 3268, pp. 597–606. Springer, Heidelberg (2004)
88. Wong, S., Pal, S.: Mining fuzzy association rules for web access case adaptation. In: Proc. of the Workshop on Soft Computing in Case-Based Reasoning (2001)
89. Zhizhen, L., Pengfei, S.: Similarity measures on intuitionistic fuzzy sets. Pattern Recognition Letter 24, 2687–2693 (2003)
90. Zhang, D., Dong, Y.: A novel web usage mining approach for search engines. Computer Networks 39(3), 303–310 (2003)
91. Zhou, B., Hui, S.C., Fong, A.C.M.: Web usage mining for semantic web personalization. In: Proc. of the Workshop on Personalization on the Semantic Web (PerSWeb 2005) (2005)
92. Xie, Y., Phoha, V.V.: Web user clustering from access log using belief function. In: Proc. of the First International Conference on Knowledge Capture (K-CAP 2001), pp. 202–208 (2001)
93. Yang, Q., Zhang, H.H.: Web-log mining for predictive web caching. IEEE Transactions on Knowledge and Data Engineering 15(4), 1050–1053 (2003)

# Chapter 5
# Web Opinion Mining and Sentimental Analysis

Edison Marrese Taylor, Cristián Rodríguez O., Juan D. Velásquez,
Goldina Ghosh, and Soumya Banerjee

**Abstract.** Web Opinion Mining (WOM) is a new concept in Web Intelligence. It embraces the problem of extracting, analyzing and aggregating web data about opinions. Studying users' opinions is relevant because through them it is possible to determine how people feel about a product or service and know how it was received by the market. In this chapter, we show an overview about what Opinion Mining is and give some approaches about how to do it. Also, we distinguish and discuss four resources from where opinions can be extracted from, analyzing in each case the main issues that could alter the mining process. One last interesting topic related to WOM and discussed in this chapter is the summarization and visualization of the WOM results. We consider these techniques to be important because they offer a real chance to understand and find a real value for a huge set of heterogeneous opinions collected. Finally, having given enough conceptual background, a practical example is presented using Twitter as a platform for Web Opinion Mining. Results show how an opinion is spread through the network and describes how users influence each other.

## 5.1 What Is Web Opinion Mining (WOM)?

On many occasions making a good decision requires the opinion of a third person, whether because of insecurity, needing a backup or not having sufficient knowledge

Edison Marrese Taylor · Cristián Rodríguez O. · Juan D. Velásquez
Web Intelligence Consortium Chile Research Centre, Department of
Industrial Engineering School of Engineering and Science, University of Chile,
Av. República 701, Santiago, Chile, P.C. 837-0720
e-mail: {emarrese,crodriguezo}@wi.dii.uchile.cl,
    jvelasqu@dii.uchile.cl

Goldina Ghosh · Soumya Banerjee
Department of Computer Science,
Birla Institute of Technology,
Mesra, India
e-mail: goldinag@gmail.com, dr.soumya@ieee.org

J.D. Velásquez et al. (Eds.): Advanced Techniques in Web Intelligence-2, SCI 452, pp. 105–126.
springerlink.com                                    © Springer-Verlag Berlin Heidelberg 2013

of the subject. One then begins to consult for information, details, comparisons and opinions in order to have a better idea on the proposal or concept at hand.

For example, wanting to buy a bike is often consulted on with other people who are more related to or have more experience on the subject, for example regarding which brand is best, what characteristics to be considered, which is more convenient -speed or mountain- and if it is better with or without shocks. After considering all the opinions given in this regard, we eventually make a decision on which bike to buy.

If the foregoing advice is considered in a business plan, it shows that for a customer to be sure about what he is going to consume, either products, services, etc., and avoid spending money needlessly or in error, it is essential to consult someone who has experience in the area. The result is the concrete idea that opinions are one of the most important indicators of personal decisions when purchasing a product, taking a tour, selecting a hotel to stay in, where to eat, etc. Many people ask their friends or family to recommend products based on their previous experiences. But there are actually more ways to communicate between persons, considering how thanks to the spread of the Internet and the continued growth of social networks like Twitter, Facebook, and other sites such as blogs or product review pages, we can now take these opinions and experiences from a bigger circle of people than just family or friends. In fact, more people check the opinions of other shoppers before buying a product, when trying to make a good decision [22] [25] [18].

Indeed, based on a survey of more than 2,000 U.S. Internet users [5], more than 75% of product review users reported that the review had had a significant influence on their purchase. Consumers reported a willingness to pay from 20% to 99% more for a 5-star-rated item than a 4-star-rated item. In another survey of 475 U.S. consumers[19], over 60% utilized on-line opinions when making purchase decision. More than 59% of consumers used the web to read on-line reviews, ratings of products or brands and research products and features, when buying products which cost between less than $100 and more than $1000.

The interest in user feedback about a product or service and the influence it has on them is very important for companies that develop products and services as well you can control how their products and their competitors' products were received by the market. As a result, you can determine what things are important to users, what features should improve, modify its advertising and many other things that can mean attracting more users to your brand.

On the other hand, views on political decisions or choices are also interesting for politicians since it allows them to evaluate how things are going, what the most important problems to be solved for the people are, whether they are likely to be elected, and so on.

For these reasons it is interesting to create a tool that can extract a set of opinions and determine what people think about certain products, services, features or be able to understand what the feelings of the people are for a politician based on the amount of positive or negative views people have on any of these topics. Depending only on the target object that has been evaluated, the term opinion mining appears in a paper by Dave et al. [6] where the ideal opinion mining tool should be to "*process a set of search results for a given item, generating a list of product attributes (quality, features, etc.) and aggregating opinions about each of them (poor, mixed, good)."*

## 5.2 How to Do WOM?

WOM is a new tool and has a long road ahead. Thus, giving a unique definition for WOM is not a simple task because the process' final objective is still unclear. It is possible to find many ways to view this problem in literature. Document Level Opinion Mining and Aspect-Based Opinion Mining are reviewed in this chapter because we consider these to be the most advanced ways to generalize a structured method to do WOM, even though a lot of other perspectives exist.

### *5.2.1 Aspect-Based Opinion Mining*

According to Bing Liu [13], opinions on the Internet can be expressed about anything, e.g., a product, a service, an individual, an organization, an event, or a topic, by any person or organization. This data from Web pages and social media could be structured text or unstructured text. The challenge is transforming the unstructured text into structured text in order to evaluate the positive, negative or neutral sentiment of opinions under study.

He defines an opinion as a quintuple $(e_i, a_{ij}, oo_{ijkl}, h_k, t_l)$ where $e_i$ is the name of an *entity* denoting the target object that has been evaluated, such as a product, service, person, event, organization or topic, and $a_{ij}$, is an *aspect* of $e_i$ meaning the components and attributes of the target object. Take for example a laptop which has a set of components, e.g. monitor, battery, CPU, and a set of attributes, e.g. size and weight. The components also have their own attributes, e.g. monitor resolution, processing capability and so on. In this model $oo_{ijkl}$ is the orientation of the opinion about aspect $a_{ij}$ of entity $e_i$, $h_k$ is the opinion holder, and $t_l$ is the time when the opinion is expressed by $h_k$. The opinion orientation $oo_{ijkl}$ can be positive, negative, or neutral or be expressed with different strength/intensity levels.

The previous definition aims to achieve the transformation of unstructured text to structured text and thereby to perform qualitative and quantitative extraction of each of the views. This quintuple gives us the specific information necessary to create a database which is easier to manage. To effect the quintuple process generation, the following tasks must be performed:

1. Extracting from an unstructured opinion, entities and their synonyms grouped in a single cluster. Each entity expression cluster indicates a unique entity $e_i$
2. Extracting the aspects associated with each of the previously-extracted entities and the grouping of those aspects in a single cluster. Each aspect expression cluster of entity $e_i$ indicates a unique aspect $a_{ij}$.

3. Extracting the authors of the opinion (holders) and the time the comment was made.
4. Finding whether the orientation of the opinion is positive, negative or neutral.
5. Creating the quintuple of each review to ensure that the entity, aspect, holder, time and orientation of opinion are well-related based on the tasks previously performed.

This is shown in the following example given by Bing Liu [13] of his approach:

*Posted by: bigXyz on Nov-4-2010: (1) I bought a Motorola phone and my girl-friend bought a Nokia phone yesterday.(2) We called each other when we got home.(3) The voice of my Moto phone was unclear, but the camera was good.(4) My girlfriend was quite happy with her phone and its sound quality.(5) I want a phone with good voice quality.(6) So I probably will not keep it.*

*Task 1 should extract the entity expressions, "Motorola", "Nokia", and "Moto", and group "Motorola" and "Moto" together, as they represent the same entity. Task 2 should extract the aspect expressions "camera", "voice", and "sound", and group "voice" and "sound" together, as they are synonyms representing the same aspect. Task 3 should find the holder of the opinions in sentence (3) to be bigXyz (the blog author) and the holder of the opinions in sentence (4) to be bigXyz's girlfriend. It should also find the time when the blog was posted, which is Nov-4-2010. Task 4 should find that sentence (3) gives a negative opinion of the voice quality of the Motorola phone but a positive opinion of its camera. Sentence (4) gives positive opinions of the Nokia phone as a whole and also its sound quality. Sentence (5) seemingly expresses a positive opinion, but it does not. To generate opinion quintuples for sentence (4), we also need to know what "her phone" is and what "its" refers to. All these are challenging problems. Task 5 should finally generate the following four opinion quintuples:*

    (Motorola, voice_quality, negative, bigXyz, Nov-4-2010)
    (Motorola, camera, positive, bigXyz, Nov-4-2010)
    (Nokia, GENERAL, positive, bigXyz's girlfriend, Nov-4-2010)
    (Nokia, voice_quality, positive, bigXyz's girlfriend, Nov-4-2010)

What makes the WOM proccess difficult, it is the fact that each of the above tasks has not yet been resolved. And to make matters worse there is some information delivered implicitly by opinions. It is still a challenge to ensure that the quintuple has a correspondence to each of its elements. But thanks to this whole process we are able to summarize the information of hundreds of thousands of opinions and determine what people feel about a product, service, etc.

Aspect-Based Opinion Mining is quite important in the opinion-mining area, since it shows an interesting way of ordering information for later analysis of large amounts of data and the ability of the quintuple to receive any other items necessary when performing a specific study. It is a methodology that somehow generalizes the process of opinion mining.

## 5.2.2   *Document Level Opinion Mining*

A different approach is proposed by [7], considering a three-phase process which is used by us to create an example of opinion mining in analyzing the influence of some users over others in the social network Twitter section 5.5.

The first phase is Corpora Acquisition Learning, whose aim is to automatically extract documents containing positive and negative opinions from the Web, for a specific domain. They propose collecting the corpus by running queries in a search engine, entering queries specifying the application domain, a seed word they want to find and the words they want to avoid (denoted by the minus "-"sign) and save a determined number of documents from the query results.

The second phase is Adjective Extraction. In relation to this task, they propose an automatic extraction of sets of relevant positive and negative adjectives. The underlying assumption is that adjectives are representative words for specifying opinions, and to achieve extraction, they apply POS (Part of Speech) tagging to recognize adjectives. They then search for associations between the adjectives contained in the documents and the seed words in the positive and negative seed sets, trying to determine whether any new adjectives are associated with the same opinion polarity as the seed words. After that, a filtering process is applied, to keep only the adjectives that are strongly correlated with the seed words. They retain rules containing more than one seed word and then consider adjectives appearing in both the positive and the negative list, applying a formula to rank adjective associations and then deleting the irrelevant ones at the end of the generated list.

The final phase is Classification of new documents using the sets of adjectives obtained in the previous phase. In order to do that, they calculate the document's positive or negative orientation by computing the difference between the number of positive and negative adjectives encountered, from both of the previously described lists. If the result is positive, the document will be classified as positive (the same is true for negative). Otherwise, the document is considered to be neutral. For this phase, it is worth mentioning a proposed method extension, which considers polarity inversion words, such as "not, neither "and "nor".

## 5.3   Sources for Web Opinion Mining

Having already defined the problem of Web Opinion Mining, the natural next question is related to the possible applications of this technique and subsequently, the set of possible web data sources to use as input. As the Web continues to grow, the number of possible sources for Web Opinion Mining grows rapidly too. Particularly, in the context of WOM, the explosive development of social media plays an important role. Within social media, it is possible to find a variety of different platforms whose content is being increasingly used by individuals and organizations for their decision making [13].

Social media includes web pages such as reviews, forum discussions, blogs, and social networks, like Facebook, Twitter, Foursquare and so on. This variety of sources suggests a heterogeneous mix of structures to work with. As a consequence of this, one needs to specify a different strategy for each source, each one oriented to the particular problem of extracting data, then processing this data and finally discovering valuable information locally within the selected source. Thus, given the mosaic of different structures and procedures for each source, the problem of designing a global methodology is complex and hasn't been widely discussed in literature. The more complete approach to this topic is the idea presented by Bing Liu, from Illinois University, in some publications and with more detail in the book *Web Data Mining*. A review of his work is presented in section 5.2.1.

The rest of this section is a review of the WOM problem in some particular sources. In the first place, the problem of extracting opinions from blogs, forums and news is studied. Then, Twitter is analyzed in order to characterize its content and determine how it can alter the Web Mining strategy to the extraction of knowledge from opinionated documents. Finally a general and brief characterization of other sources is presented.

### 5.3.1 Blogs, News and Forums

The sources that are probably the richest in opinionated documents are blogs, news and forums. These sources present some common features that make them a good choice when deciding where to look for opinions, but before any analysis, it is worth mentioning that there are some differences between these sources. News and blogs are two important sources of opinions. The writing of the former is formal in comparison to that of the latter since blog articles expressing personal opinions are often written in casual style. Because of this, generally speaking, news documents are more objective, while blog articles are usually more subjective. Other important differences refer to opinion holders, which probably belong to different social classes. Opinions extracted from news are mostly from well-known people, while opinions expressed in blogs may come from a "no name". This issue turns out to be even more important in forums, where it is often difficult to determine the real opinion holder's name [12].However, when analyzing a specific public issue, listing opinions from different sources in parallel can provide many views of the issue, which helps to understand how different social actors react toward the same situation. On the other hand, the most important common feature between news and blogs refers to the document extension, which compared with forums and other sources is much longer. At first glance, this could be seen as an advantage, but long document extensions present a new main problem in Opinion Mining, i.e., how to determine where an opinion sentence is? To solve this problem, major topic-detection techniques are proposed in [12], [7], and some other related publications, to capture main concepts embedded implicitly in a relevant document set. In relation to forums, document extensions can be quite different among singular topics or communities, so it is difficult to establish a main tendency within this field. However, a useful and positive feature that appears in a high number of review forums is the post-rating system. Post ratings can be used as a predictor of the content's semantic orientation or to contrast and validate text processing and analysis results.

## 5.3.2  Twitter

Twitter is often considered a microblogging platform, but it is also frequently included as a social network. Given the features of this platform, probably both of these considerations have a certain degree of truth, but for the problem of Web Opinion Mining, the fact that Twitter is a microblog is highly relevant. Microblogging is a growing popular communication channel on the Internet, where users can write short text entrances in a public or private way (to a group of contacts). Messages are extremely short, allowing users to write a maximum of 140 characters on each post, called a tweet. These tweets can be written through the Twitter web interface or through a variety of mobile devices, like smartphones, some cell phones and other devices. These short messages can be seen as being a newspaper headline and subtitle, which makes them easy to produce (write) and process (read). This feature makes microblogs unique when compared to other similar web communication media, like blogs and web pages [2]. As a microblogging social network, the first relevant feature of Twitter's messages are their brevity, which make users look for various special ways to add content in the messages. The most-used approaches are adding content indirectly or trying to use fewer characters to express the same ideas. One important fact on the first topic is the inclusion of links to other web sites to indirectly complement the content of the tweet. In addition to this, it is possible to find a high density of messages containing short URL services, which in fact were created to help Twitter users to include these links with a low number of characters. Another issue refers to the use of Twitter's special characters, like hashtag (#) to denote a particular topic or to call a user, and RT, short for retweet. On the other hand, a different problem associated with tweets is that they are written in colloquial or informal language. In addition to the brevity of messages, this supposes the use of many colloquial symbols or expressions that, in order to be understood as regular text, require preprocessing. In relation to this issue, it is possible to find four different main features:

1. One first special aspect of tweets is that they often include a variety of emoticons. These emoticons help users to express their ideas, feelings or moods in fewer characters, but have a deep impact on text processing. As emoticons are not considered words, they do not have any structure that helps in extracting their lexical meaning, and are not formally included in any dictionary or language that helps to understand their meaning. Nevertheless, as proposed in [20], emoticons can be successfully used to previously determine the tweet's sentiment orientation, opening a wide new field of investigation. Based on this proposal, in [16] emoticons are used to create a corpus collection strategy, defining

two kinds: happy (including "=)", ":)",":D"and others similar) or sad (":-(",
":(", "=(", etc.)

2. In the second place, spelling mistakes are also common in tweets. This implies
the implementation of a preprocessing task in order to fix possible mistakes and
ensure a correct interpretation of the content. In relation to this, [11] proposes
a technique based on the Levenshtein algorithm, which determines a notion of
distance between the misspelled and actually-meant word. This can be used
combined with a dictionary to rank and then select the most probable letter
replacements from a list of previously-generated possible word candidates.
3. Another feature commonly found in tweets is the repetition of one letter in a
word. This is mostly done when users want to add some emotive intensification
to the text, for instance repeating vowel letters as in "I loooove you". In this
field, [1] proposes a control system based on regular expressions for Spanish
with back reference, replacing the appearance of two or more characters by
only one letter, excepting the groups "cc", "ll"and "rr", which are commonly
used in this language.
4. A last feature is related to the use of Internet language, or "Netspeak", in the
messages. In this context, the use of capital letters can be problematic when
tokenizing or lemmatizing text during preprocessing.

Finally, it is interesting to see Twitter as a network of related users, who share
opinions among themselves and influence each other. Twitter provides some useful
tools that can be used to analyze how a specific topic or opinion tendency is spread
through the network, and discover users who influence others or are more easily influenced by another. Based on this, one could be able to, for instance, define clusters
and find non-trivial segmentations based on the characterization. A proposal on this
field is presented in section 5.5.

### 5.3.3   Other Media

There are a lot of other possible sources for WOM. In the context of social networks,
Facebook is often proposed as a powerful and complete repository. Nevertheless, the
main problem in this case is related to privacy policies and access limitations to the
contents. Thus, opinionated documents are not always publicly available, and it is
difficult to reach them.

## 5.4   Visualization Techniques

In most of the cases, studies need to analyze a large number of opinion holders. Common sense indicates that one opinion from a single holder is usually not
sufficient for action. This idea naturally leads to the task of opinion summarization. The literature proposes some different approaches to summarizing and then

visualizing summarized opinions. This section focuses on different visualization techniques which require, in one way or another, some kind of previous summarization. This last task is a complex and pretty well-studied field. Its application to opinions (and web opinions) is just a particular case and will be briefly discussed in this section.

As presented in [3], traditional summarization consists of constructing new sentences from the opinionated document, in order to extract the document main points. In [3], the opinion summarization technique proposed is founded on the idea of analyzing relationships among basic opinionated units within the document. More precisely, the paper presents an approach to multi-perspective question answering (MPQA) that views the task as one of opinion-oriented information extraction. Briefly, the information extraction system takes as input an unrestricted text and summarizes the text with respect to a previously-specified topic or domain of interest, finding useful information about the domain and encoding that information in a structured form, suitable for populating databases. The process involves creating low-level annotations of the text which are then used to build the summary. Visualization in this context is thus the construction and presentation of short sentences (or sets of sentences) that capture a document's main opinionated ideas.

The traditional fashion for summarization then means producing a short text summary that gives the reader a quick overview of what people think about a defined object. Some traditional summarization techniques can be found in [4], [15], [21] and [23]. Nevertheless, the main weakness of these text-based summaries is that they are just qualitative, which means that it is not possible to apply any numerical or quantitative analysis to them. As proposed in [13], the quantitative side is crucial, just as in traditional survey research.

In this context, opinion quintuples defined by Liu's approach are a good source of information for generating both qualitative and quantitative summaries, and can be stored in database tables. Based on this, a kind of summary based on aspects is defined, which is called aspect-based opinion summary [9], [10]. Having built the proposed structure, a whole set of visualization tools can be applied to see the results in all kinds of ways, to then gain insights into the opinions. In this case, bar charts or pie charts are both used. As an example, data can be visualized using a bar chart in which each bar above the X-axis shows the number of positive opinions on one aspect, and the corresponding bar below the X-axis shows the number of negative opinions on the same aspect. A different technique may only consider showing the percent of positive opinions.

Liu's visualization proposal is also interesting because it enables comparison of opinion summaries of some competing products. In addition to this, instead of just generating a quantitative summarization, a text summary directly from input reviews is also possible, generating natural language sentences based on what is shown in the charts [14].

It is important to mention that this technique is only related to product opinions and results quite differently from traditional text summarization because it focuses and mines only the features of these products, while also determining whether the opinions are positive or negative. There is no rewriting of original sentences to capture the main points of the opinionated selected document, as in the classic text summarization described previously.

On the other hand, a completely different approach in relation to text summarization and visualization is presented in [12]. As presented before, traditional summarization algorithms rely on the important facts of opinionated documents and remove redundant information. Nevertheless, it is likely that sentiment degree and correlated events play major roles in summarization. Because of that, [12] proposes that repeated opinions of the same polarity cannot be dropped because they strengthen the sentiment degree, but repeated reasons why they stated a position should be removed when generating summarization. To apply this summarization system it was therefore needed to know which sentences were opinionated and then decide if they focused on a designated topic. An algorithm that detects and extracts major topics in long documents and then classifies them in positive or negative orientation in relation to that topic was then developed. Then, for brief summarization, the document with the largest number of positive or negative sentences is picked up and its headline is used to represent the overall tendency of positive-topical or negative-topical sentences. For detailed summarization, a list of positive-topical and negative-topical sentences with higher sentiment degree is generated.

As a complement, an opinion-tracking system that shows how opinions change over time is proposed. The tracking system is very similar to Liu's proposal and consists of bar charts that simply count the number of positive-topical and negative-topical sentences on a selected topic at different time intervals. Nevertheless, a large number of relevant articles is required.

Finally, Tateishi's approach is worth mentioning, which introduces radar charts for summarizing opinions and has been frequently cited in the literature. A more detailed description of this technique can be found in [24]. Sadly, as the original document is only available in Japanese as this text is being written, it was not possible to include a deeper analysis due to language issues.

## 5.5   An Application of WOM in Twitter

This example envisages the social network Twitter. The record set provides us with some vital information such as user ID, number of followers, number of following users, number of tweets and frequency of tweets. There can be both important and unimportant information that can be shared. This information can be transmitted through the blogs where some senders take up the initiative to transform different kinds of events. Based on the blogs the person can have followers. Followers are those who either agree with an opinion or get influenced by other's events and propagate these messages to yet others. Again, if we consider the number of blogs or tweets and their values, then the number of influential members of the tweeter social network can also be identified. Based on the tweeter database it is possible to cluster the number of followers and following individuals. By applying the subtractive clustering method in fuzzy logic the grouping of those members who are more influential is done.

## 5.5.1   Extracting Significant Information from Tweets

In this section a methodology for extracting and processing tweets is proposed by using *Subtractive Clustering using fuzzy logic* with the application of logical operations and the *if-then rule* statement. The relevant terminologies are as follows:

- **Fuzzy Inference System:** This is the process of formulating the mapping from a given input to an output using fuzzy logic. The mapping process can then be applied to make decisions or perform any pattern analysis. The Fuzzy Inference System also involves the concepts of membership, logical operations and the if-then rule.
- **Membership functions:** Membership functions are curves that define how each point in the input space is mapped to the degree of membership between 0 and 1. The input space is also called 'Universe of Discourse'.
- *Logical operations* like AND, OR, NOT are also applied.
- 'If-then rule'statements are used for conditional statements that are comprised of fuzzy logic.
- *Subtractive Clustering:* This method of clustering functions by assuming each data point to be a potential cluster center and calculates a measure of the likelihood that each data point would define the cluster center based on the density of the surrounding data points. Subtractive clustering is a fast, one-pass algorithm for estimating the number of clusters and the cluster centers in a set of data. The cluster estimates obtained can be used to initialize iterative optimization-based clustering methods and model-identification methods. The steps in performing the process are as follows:

  - Selecting the data point with the highest potential to be the first cluster center.
  - Removing all data points in the vicinity of the first cluster, called a *radii*, in order to determine the next data cluster and its center location
  - Repeating this process until all of the data are within the *radii* of a cluster center.

The *radii* are a vector of entries between 0 and 1 that specify a cluster center's range of influence in each of the data dimensions. The best value of a given *radii* is usually between 0.2 and 0.5.

A *Sugeno-type fuzzy inference* system that models the data behavior is generated to provide a fast, one-pass method to take input-output training data. A typical rule in a *Sugeno fuzzy* model has the form,

'If Input 1 = $x$ and Input 2 = $y$, then Output is $z = ax + by + c$'.

For a zero-order Sugeno model, the output level $z$ is a constant ($a = b = 0$). The output level $z_i$ of each rule is weighted by the firing strength $w_i$ of the rule. The final output of the system is the weighted average of all rule outputs, computed as:

$$FinalOut = \frac{\sum_{i=1}^{N} w_i z_i}{\sum_{i=1}^{N} w_i} \qquad (5.1)$$

where $N$ is the number of rules.

## 5.5.2 Data Quality Evaluation

The first step in any information extraction process is the evaluation of the data quality to be processed. In the case of text originated in Twitter, these can content many wrong words, typos mistakes, emoticons[1], non structured sentences, etc. A second step is to select the algorithms to use for processing the Twitter data and finally, a mathematical treatments of the extracted data for detecting the most influential members of the Twitter community.

### 5.5.2.1 Components of Standard Data Set

To validate the proposed analytical solution, we collect the data from the Twitter social networking site. The database is extracted using the three main streaming products: The Streaming API, User Streams and Site Streams.

*Streaming API:* It is used to collect public statuses from all users, filtered in various ways. It can be by user ID, by keyword, by random sampling, by geographic location and other parameters.

*User Streams:* All the data are required to be updated. It provides public and protected statuses from followings, direct messages, mentions, and other events taken on and by the user.

*Site Streams:* This allows multiplexing of multiple User Streams over a Site Stream connection.

While using the Search API one is not restricted by a certain number of API requests per hour, but instead by the complexity and frequency. As requests to the Search API are anonymous, the rate limit is measured against the requesting client IP. The record set collected contains certain information about 99 unique members of the Twitter community where each member's 'number of followers', 'number of following', 'number of tweets'and 'frequency of tweets'are present. Table 5.1 shows the details of the record set:

---

[1] An emotion can be defined as "feeling states with physiological, cognitive, and behavioral components" [8], in that sense an emoticon can be defined as a short sequence of keyboard letters and symbols for expressing a human emotion in chats, blogs, e-mails, twitter, etc.

**Table 5.1** The detailed record set of Twitter

| user_id | no_of_followers | no_following | no_of_tweets | tweet_frequency |
|---|---|---|---|---|
| 156518278 | 66 | 80 | 4248 | 48.116505 |
| 33765767 | 682 | 660 | 11106 | 74.75192 |
| 450587971 | 32 | 65 | 297 | 36.473682 |
| 380242792 | 7 | 0 | 445 | 20.493422 |
| 20531842 | 279 | 392 | 13876 | 87.58521 |
| 55989455 | 611 | 337 | 51513 | 376.39978 |
| 83018127 | 0 | 0 | 38 | 0.3089431 |
| 224300080 | 61 | 1 | 55886 | 883.0745 |
| 27937504 | 1285 | 465 | 12547 | 82.857544 |
| 467461925 | 31 | 0 | 1331 | 251.81082 |
| 467464356 | 21 | 0 | 1112 | 210.37839 |
| 163777200 | 587 | 439 | 23128 | 270.7291 |
| 402424165 | 2 | 0 | 298 | 17.98276 |
| 180749032 | 141 | 145 | 9097 | 114.94404 |
| 18370866 | 465 | 190 | 6554 | 39.65255 |
| 14115513 | 125 | 25 | 547 | 2.6479945 |
| 144538924 | 1034 | 244 | 14456 | 155.91988 |
| 20155063 | 165 | 117 | 1519 | 9.544884 |
| 203226606 | 100 | 0 | 26745 | 376.69016 |
| 102953286 | 191 | 195 | 6766 | 60.876606 |
| 132729891 | 2805 | 117 | 1265 | 12.9838705 |
| 214006837 | 458 | 373 | 20129 | 298.5233 |
| 221106075 | 32 | 71 | 4446 | 68.85398 |
| 224996941 | 127 | 109 | 5610 | 88.84615 |
| 40540088 | 170 | 286 | 1916 | 13.226824 |
| 330240789 | 120 | 204 | 1453 | 43.465813 |
| 465856538 | 3 | 9 | 16 | 2.8717947 |
| 430139433 | 2 | 14 | 178 | 15.575 |
| 93816436 | 624 | 396 | 26756 | 229.52452 |
| 467452745 | 32 | 0 | 961 | 181.81082 |
| 281369631 | 32 | 42 | 228 | 5.018868 |
| 76161854 | 0 | 0 | 49 | 0.38713318 |
| 124483151 | 7 | 20 | 25 | 0.24752475 |
| 401097105 | 1 | 0 | 29 | 1.720339 |
| 234992003 | 1043 | 864 | 42475 | 718.17633 |
| 80854705 | 902 | 1936 | 782 | 6.2991943 |
| 348272136 | 957 | 1549 | 146 | 4.985366 |
| 138510278 | 124 | 159 | 2909 | 30.575073 |
| 274335070 | 6 | 7 | 720 | 15.180723 |
| 108578780 | 386 | 383 | 1249 | 11.503947 |
| 289726223 | 522 | 10 | 18871 | 437.4073 |
| 213000229 | 631 | 0 | 10536 | 155.59494 |
| 456494008 | 28 | 1 | 1409 | 197.26 |
| 468635975 | 0 | 11 | 4 | 0.7777778 |
| 213000229 | 631 | 0 | 10536 | 155.59494 |

### 5.5.2.2 Processing Data Originated in Twitter

To design the data processing stage, the following steps need to be accomplished:

A Comparison among the 'number of followers'and 'number of following'data of each member to find out which ones all are influential and which ones all are influenced. The condition is that if number of followers > number of following, then a member is categorized as being influential and if not, then as being influenced.
B Based on the number of tweets the tweet weight age is judged by providing a threshold value and then analyzing whether the numbers of followers are directly proportional to the number of tweets made by each member.
C If the numbers of tweets along with their weighted value are high, then those members can have larger number of followers. This measurement is done by checking the direct proportionality among the number of tweets and number of followers. If it is inversely proportional then the number of tweets was not influential.
D An analysis report that is derived from the above two conditions will help in clustering those members who are more influential than the others based on the data set of Twitter in Table 5.1 by using the subtractive clustering technique in fuzzy logic.

### 5.5.2.3 Modeling the Twitter User Behavior

Certain information is available from the Twitter database, including each individual having an ID number, along with their number of followers and following. Based on these two fields, an analytical report of the person's nature can be identified to determine whether the person is influential or influenced. Based on these data a very simple analysis can be derived that is a symmetric relation. Symmetric relation can be expressed as *"if x is related by R to y, then y is related by R to x"*.

Here, the numbers of individuals (unique members of the Twitter community) are denoted as,

$$I = \{I_1, I_2, I_3, ..., I_n\}$$

and the nature of the individual can be categorized in the following form,

$$Nature = \{Influencial, Influenced\}$$

By applying the concept of symmetric relation the inference that can be drawn is: If number of followers > number of following, then member is categorized to be influential and if not then influenced.

Analyzing fields which provide information about the number of tweets made by each member can help in judging the rate of increase or decrease in followers. The number of followers will be directly proportional to the number of tweets sent, if the tweets are larger in number and also the weightage of each tweet is high, which means the message is important and valuable. This property can be represented in the following format:

$$NF_i = NT_j * k,$$
$$i = \{i_1, i_2, ..., i_n\}$$
$$j = \{j_1, j_2, ..., j_m\}$$

(5.2)

where $NF$ = Number of Followers, $NT$ = Number of Tweets, $k$ = Constant.

The reverse representation happens when the numbers of tweets are less in number and when they are of less importance. This means that the number of followers will be inversely proportional to the number of tweets. This property is represented in the following way:

$$NF_i = \frac{k}{NT_j},$$
$$i = \{i_1, i_2, ..., i_n\}$$
$$j = \{j_1, j_2, ..., j_m\}$$

(5.3)

Where, NF= Number of Followers, NT= Number of Tweets, k= Constant

Thus from equations 5.2 and 5.3, we can detect the most influential members of the tweeter community based on the database of the particular moment by applying subtractive clustering. The member with the highest potential is selected to detect the first cluster, and then all the points are removed from the vicinity of the first cluster so as to detect the next cluster. This process is repeated again and again so as to make a collection of all the members that are influential as well as whether their tweets are important or of high value. The data behavior that can thus be identified from the record set is a one-pass method to take input-output training data and generate a Sugeno-type fuzzy inference system.

## 5.5.3   Analysis and Results

The effectiveness of measuring the different levels of followers is determined in Figures 5.1 and 5.2. In Figure 5.1 the different rankings of members are plotted based on the number of followers less than the number of following. The x-axis denotes the different number of followers and y-axis denotes the different ranking of the followers based on the different tweet weight age.

In figure 5.2 the different rankings of members are plotted based on the number of followers greater than following.

Depending on the number of tweets and also the valuation of those tweets the number of followers is denoted. Figure 5.3 and figure 5.4 show the scattered plotting of those members whose tweets were important and collected a larger number of followers whereas the members who had less important tweets had a lesser number

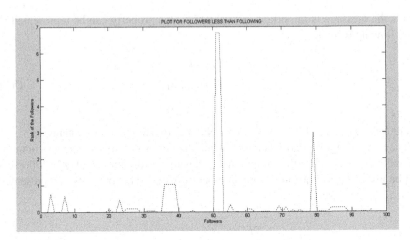

**Fig. 5.1** Members with followers less than following

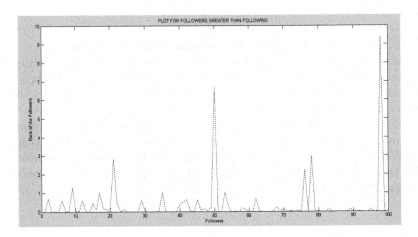

**Fig. 5.2** Members with followers greater than following

of followers. Figure 5.3 describes the collection of those members who had higher numbers of followers due to their strength in tweets.

Similarly, figure 5.4 denotes the reverse condition where the numbers of followers are less due to the less important tweets.

Hence, based on these results, we can draw a conclusion about those members who are very influential and shared a lead role. The following result is plotted in figure 5.5 by applying the subtractive clustering technique.

The social network is a place that helps in sharing different opinions with different users. In this way plenty of information could be provided about knowledge being shared. Additionally, in those cases where some vital tweets are made by some individual then those tweets are given much more importance than the

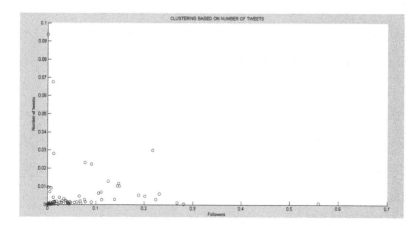

**Fig. 5.3** Important tweets with influenced followers

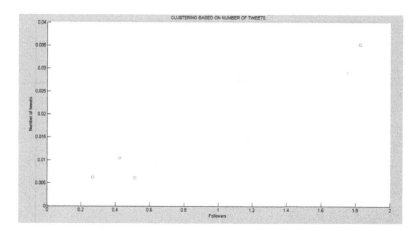

**Fig. 5.4** Unimportant tweets with less influenced followers

others, since they get propagated among more followers. This can be represented in a graphical representation where the x-axis denotes the number of individuals transferring tweets and the y-axis represents the flow of the tweets. This scattered graph denotes when the followers rate was high due to the high weight age of the tweet. The number of followers rate increases based on time, event and the members of the tweeter group. The rate of increase in flow of the tweets among the individuals can also be expressed mathematically, $y = x^a$, where $x$ = weight of the tweet (high-valued tweets will be transmitted to many individuals) $y$ = number of followers following that tweet and transferring them to others (depending on the tweet value) $a$ = depends on the previous value of $y$.

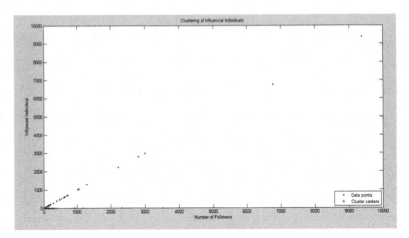

**Fig. 5.5** Subtractive clustering of all influential members

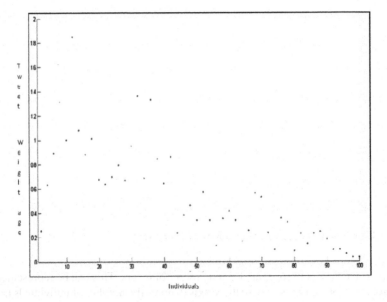

**Fig. 5.6** Scattered plot denoting flow of influential tweets

The most influential person of a group can always be ranked as the best informer of some important event, news or for sharing good thoughts. In Figure 5.7 bar diagram denotes here the best member, who had communicated well and had been placed in the high rank. This is followed by the ranks of the other members' as per their weight of the tweet. The x-axis denotes the individual members and the y-axis represents the rank level of each individual where they belong.

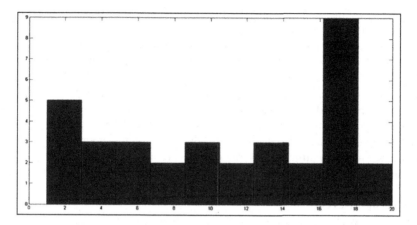

**Fig. 5.7** Histogram denoting rate of flow of influential tweets

### 5.5.4   Tweeter as Novel Web User Behavior Data Source

In fact, the data originated in social networks are new sources for understanding the web user behavior. A very important issue in this kind of data is these normally are generated by own web user decision, corresponding usually what the web user is thinking and feeling, However, these opinion can be influence by other web user opinions, then to analyze what influenceable a social network can became will be always necessary for analyzing the web user behavior.

A social network community like Twitter, the followers receive the others web user influence by tweets an retweets opinions about some particular theme. Then this kind of social site is not only a medium of sharing important events but also in taking part in knowledge sharing and also participating in some conversations. The blogs that are posted by the members of such social network community are some time a mind opener for other members also. This leads in judging whether the blogs posted by the person is valuable which further leads in categorizing the influenced and influencing persons. This can be finally subjected towards many real life applications as sentiment analysis and emotion modeling pertaining to social network analysis.

## 5.6   Summary

With the inception of Web 2.0 and the explosive growth of social media on the Web, users have now the possibility to express personal opinions about products, services, and therefore, access a huge repository of these opinions to make better decisions about buying or using a product or service. This information could also be important for enterprises, that might be interested in knowing how their products or services are perceived by the market.

Among the different sources where users add their personal opinions, we distinguished and discussed four: Blogs, News, Forums and the social network Twitter. Each one presents a specific operating mode which brings particular issues related. In section 5.3 we analyzed these four sources and tried to determine how their particular problems can affect the mining process.

Having given a look at all these issues, WOM begins to become a complex and challenging task. Then we stated that developing a system that embraces all these problems is naturally difficult. Nevertheless, it is possible to find various approaches in literature. In section 5.2 we introduce two of these methodologies: Document Level Opinion Mining and Aspect-Based Opinion Mining, even though a lot of other perspectives exist.

Each approach proposes a series of logical steps to transform the complex and heterogeneous data from different sources into a structured and simpler configuration. Nevertheless, giving a unique definition for WOM is not a simple task because the process final objective is still unclear.

On the other hand, in section 5.4 we presented summarization and visualization techniques that allow a better comprehension of the mining process results. For this reason these are remarkably important tasks and deserve to be studied. Multiple techniques involving classic text summaries, chart visualization methods and other new proposals were introduced.

Finally, in order to achieve a better understanding of one Web Opinion Mining existing technique, we presented a practical example in section 5.5, which implements Document Level Opinion Mining in the social network Twitter. In this case, the goal was detecting the most influential users on Twitter community.

# References

1. Balbachan, F., Dell'Era, D.: Automatic sentiment analysis of short texts in twitter platform (2011) (in Spanish)
2. Bifet, A., Frank, E.: Sentiment Knowledge Discovery in Twitter Streaming Data. In: Pfahringer, B., Holmes, G., Hoffmann, A. (eds.) DS 2010. LNCS, vol. 6332, pp. 1–15. Springer, Heidelberg (2010)
3. Cardie, C., Wiebe, J., Wilson, T., Litman, D.: Combining low-level and summary representations of opinions for multi-perspective question answering. In: Proceedings of the AAAI Spring Symposium on New Directions in Question Answering, pp. 20–27 (2003)
4. Carenini, G., Ng, R., Pauls, A.: Multi-document summarization of evaluative text. In: Proceedings of the European Chapter of the Association for Computational Linguistics (EACL), pp. 305–312 (2006)
5. comScore/the Kelsey group. Online consumer-generated reviews have significant impact on offline purchase behavior. Press Release (November 2007),
http://www.comscore.com/press/release.asp?press=1928

6. Dave, K., Lawrence, S., Pennock, D.M.: Mining the peanut gallery: Opinion extraction and semantic classification of product reviews. In: Proceedings of the 12th International Conference on World Wide Web, pp. 519–528. ACM (2003)

7. Harb, A., Plantié, M., Dray, G., Roche, M., Trousset, F., Poncelet, P.: Web opinion mining: How to extract opinions from blogs? In: Proceedings of the 5th International Conference on Soft Computing as Transdisciplinary Science and Technology, pp. 211–217. ACM (2008)

8. Hsee, C., Hatfield, E., Carlson, J.G., Chemtob, C.: The effect of power on susceptibility to emotional contagion. Cognition and Emotion 4, 327–340 (2004)

9. Hu, M., Liu, B.: Mining and summarizing customer reviews. In: Proceedings of the Tenth ACM SIGKDD International Conference on Knowledge Discovery and Data Mining, pp. 168–177. ACM (2004)

10. Hu, M., Liu, B.: Opinion extraction and summarization on the web. In: Proceedings of the National Conference on Artificial Intelligence, vol. 21, p. 1621. AAAI Press, Menlo Park (2006)

11. Jurafsky, D., Martin, J.H., Kehler, A., Vander Linden, K., Ward, N.: Speech and language processing: An introduction to natural language processing, computational linguistics, and speech recognition, vol. 2. Prentice-Hall, New Jersey (2000)

12. Ku, L.W., Liang, Y.T., Chen, H.H.: Opinion extraction, summarization and tracking in news and blog corpora. In: Proceedings of AAAI-2006 Spring Symposium on Computational Approaches to Analyzing Weblogs, vol. 2001 (2006)

13. Liu, B.: Web data mining: exploring hyperlinks, contents, and usage data. Springer (2011)

14. Liu, B., Hu, M., Cheng, J.: Opinion observer: analyzing and comparing opinions on the web. In: Proceedings of the 14th International Conference on World Wide Web, pp. 342–351. ACM (2005)

15. Nishikawa, H., Hasegawa, T., Matsuo, Y., Kikui, G.: Optimizing informativeness and readability for sentiment summarization. In: Proceedings of the ACL 2010 Conference Short Papers, pp. 325–330. Association for Computational Linguistics (2010)

16. Pak, A., Paroubek, P.: Twitter as a corpus for sentiment analysis and opinion mining. In: Proceedings of LREC (2010)

17. Pang, B., Lee, L.: Opinion mining and sentiment analysis. Found. Trends Inf. Retr. 2(1-2), 1–135 (2008)

18. Park, D.H., Kim, S.: The effects of consumer knowledge on message processing of electronic word-of-mouth via online consumer reviews. Electronic Commerce Research and Applications 7(4), 399–410 (2009)

19. Razorfish. Digital consumer behavior study (2007)

20. Read, J.: Using emoticons to reduce dependency in machine learning techniques for sentiment classification. In: Proceedings of the ACL Student Research Workshop, pp. 43–48. Association for Computational Linguistics (2005)

21. Seki, Y., Eguchi, K., Kando, N., Aono, M.: Opinion-focused summarization and its analysis at duc 2006. In: Proceedings of the Document Understanding Conference (DUC), pp. 122–130 (2006)

22. Shin, H.S., Hanssens, D.M., Kim, K.I., Gajula, B.: Impact of positive vs. negative e-sentiment on daily market value of high-tech products (2011)

23. Stoyanov, V., Cardie, C.: Partially supervised coreference resolution for opinion summarization through structured rule learning. In: Proceedings of the 2006 Conference on Empirical Methods in Natural Language Processing, pp. 336–344. Association for Computational Linguistics (2006)
24. Tateishi, K., Fukushima, T., Kobayashi, N., Takahashi, T., Fujita, A., Inui, K., Matsumoto, Y.: Web opinion extraction and summarization based on viewpoints of products (in japanese). Information Processing Society of Japan SIGNL Note 93, 1–8 (2004) (in Japanese)
25. Zhu, F., Zhang, X.: Impact of online consumer reviews on sales: The moderating role of product and consumer characteristics. Journal of Marketing 74(2), 133–148 (2010)

# Chapter 6
# Web Usage Based Adaptive Systems

Pablo Loyola Heufemann, Jorge Gaete Villegas, and In-Young Ko

**Abstract.** The Internet is becoming an important tool for the realization of day-to-day activities, which leads to a new level of interaction between users and software systems. This new scenario presents endless opportunities as well as enormous challenges. In order to tackle these, user-adaptive software systems have been recently used. These technologies aim to allow computer systems to dynamically modify their content, structure and presentation for better delivery of the available resources, while considering the user's interest and behavior, and most recently, mobile environments. This chapter overviews the newest technologies in the area of user-adaptive software systems applied to Web environments and proposes a set of directions for the future development of Web Usage Based Adaptive Systems in the new Internet environments.

## 6.1 Introduction

Web Usage Based Adaptive Systems aim to allow computer systems to dynamically modify their interfaces for better presentation of the available content in

Pablo Loyola Heufemann
Division of Web Science and Technology, KAIST
335 Science Road, Yuseong-gu Daejeon 305-701 Republic of Korea
e-mail: ployola@kaist.ac.kr

Jorge Gaete Villegas
Department of Computer Science, KAIST,
291, Daehak-ro, Yuseong-gu, Daejeon 305-701, Republic of Korea
e-mail: jorge@kaist.ac.kr

In-Young Ko
Department of Computer Science and Division of Web Science and Technology, KAIST
e-mail: iko@kaist.ac.kr

J.D. Velásquez et al. (Eds.): Advanced Techniques in Web Intelligence-2, SCI 452, pp. 127–148.
springerlink.com © Springer-Verlag Berlin Heidelberg 2013

consideration of the user's interest, past behavior, and abilities [2] as well as, most recently, in consideration of mobile environments [4], and human-related processing capabilities [9].

Adaptation is presented in the literature as the output of a clearly characterized process described by its inputs, outputs and transformation methods [25]. Adaptation, then, is the outcome of the interaction between three models taking into account the users and the system's resources and transformation strategies and functions. In the following chapter, this process will be explained by describing these models, namely the Domain Model (system's resources), User Model and Transformation Model.

The second section of this chapter overviews these models by describing them and by presenting the interaction between them. The third section focuses on how adaptation is accomplished, by providing the state-of-the-art in adaptation techniques. Finally, given the importance of privacy issues in the future internet context, a legal issues section has been added in order to give a wider understanding of all the aspects involved when personalization is intended.

### 6.1.1  An Overview on Web Usage Based Adaptive Systems

As Web technology allows companies to exploit the advantages of providing services and aiding customers via the Internet, common users have closer and more frequent interaction with software systems. This represents endless opportunities as well as enormous challenges.

From the system's point of view, the challenges are related with the issue of supporting the heterogeneous needs of many users. From the user's point of view, the speed at which new information is being produced and services offered can be overwhelming. Complex software systems are being developed to avoid the limitations of the "one-size-fits-all" [9] approach and to effectively support users in their activities by taking advantage of the resources available on the web.

We refer to such software systems as "Web Usage Based Adaptive System". Such systems are defined in this chapter as software systems in the web domain that are able to modify their structure and content based on how users interact with them. The two major features of these systems are user internalization into the system and the use of resources and data available on the web.

The user is taken into account by a user model in the system design, while the source of resources are the Internet, web browsers and the Internet of things. Web Usage Based Adaptive System usage is extensible to any software system that needs to interact with users on the web, and for which personalization is required. Web Usage Based Adaptive System usage can be found in multiple areas as e-learning, e-commerce, community websites, tourism, medicine, and information retrieval, etc.

## 6.1.2 Web Usage Based Adaptive Systems Evolution

Software systems respond to the needs of users, and therefore the more complex these needs are, the more complex the systems will be. The rapid development of the Internet, the richness of resources on the web, and the variety of users propels the development of today's sophisticated software systems.

User' Adaptive Software Systems can be traced back to early 1990's. At that stage, the work was focused on adaptive interfaces and hypertext study in closed corpus systems. This first generation of Adaptive Systems was a "pre-Web" generation. The work focused on adapting the presentation and navigation of a closed corpus system's resources as well as modeling the user's knowledge of the domain defined in particular systems [7].

Driven by the complexity of and growing amount of information available on the Internet, the fields of Adaptive Hypermedia (AH) and Web Systems (WS) merged. This second generation of Adaptive Systems [3] leveraged the Internet and, by merging user-adaptive systems and the new existing content (hypertext, multimedia, etc.), created an adaptation effect. Since the system was no longer closed, resources could be included dynamically and it was possible to include adaptive content in consideration of not only a user's knowledge but the user's interests as well.

The current third generation of adaptive systems is mobile based, and context is being taken into consideration. These systems apply the idea of adaptation to the user's surrounding conditions, to adapt not only to the user but to the user's context as well [11, 38].

This history of development makes it clear that software systems respond to technological advances. In this sense, the advent of the Internet of things and the Internet of services [30] opens new frontiers for Web usage based adaptive systems. Including not only information on the user and the context, but also information on available services and resources will foster the development of new interactions and strategies for aiding the user.

## 6.2 Process Framework

In this section, the general process by which adaptation takes place is described. By presenting this general process we aim to summarize the work realized particularly in the Adaptive Hypermedia [9] field and update it to include the newest technologies available.

Adaptation of a system can be seen as a process defined by the interrelation of 3 major components to respond to 3 basic questions [25, 49]:

- What to adapt? This is answered by modeling the system's resources and information structure. This model represents the Domain Model (DM) in existing literature, and its objective is to determine the environment where the user will interact with the system.

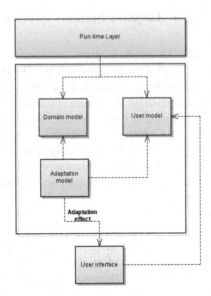

**Fig. 6.1** Models interaction. Adapted from [25, 9].

- What to adapt to? The goal of the adaptation process is defined by understanding the user. The User Model (UM) represents the user as a part of the system, by describing the user in terms of the user's characteristics, preferences and behavior.
- How to adapt? This is solved by means of the Adaptation Model (AD). The purpose of this model is to specify how the different elements of the DM are to be organized given a specified user (UM). This AM may be understood as a "teaching model", "narrative model", or any model accounting for the dynamic relationship existing in human interactions.

This division of concepts allows better understanding of the adaptation process by clearly dividing the three main questions needed for effectively supporting the user.

As shown in Figure 1, the adaptation model adapts the systems presentation by altering the available elements from the domain model. This is done by considering the user information in the UM. Since the system allows the user on-line interaction, the domain may be altered by the user as well. The system itself is feed on run time by the usage of sensors, user behaviour mining software or user manipulation [25, 9].

In this section, an overview of these main models will be presented. The next section will deal with the Adaptation model in greater detail, and is the main focus of this chapter.

## 6.2.1   Domain Model

The Domain Model (DM) describes an application domain and links this representation with the resources available in the system. The DM defines the object of manipulation by mapping the application domain into concepts. These concepts, defined in a computable representation, and the relationships between these concepts define what to modify for the adaptation [4].

### 6.2.1.1   Domain Model Elements

Elements in the DM are concepts encapsulating information from the application domain. These concepts are defined as the abstraction of information fragments that can be combined to create more complex abstractions.

In general [49], there are two logical levels of concepts: i) atomic concepts, and ii) composite components. While atomic concepts are information fragments meaningless by themselves, composite concepts are meaningful information pieces in the application domain.

The actual representation varies depending on the application domain and the goal supported by the Web usage based adaptive systems. These concepts can appear represented as fragments, pages [4], chapters, information units [31] and, in general terms, any mechanism to encapsulate information about a concept.

In concrete terms, a concept can be computed as a tuple (attribute, value) [4], as classes or methods [31], as metadata associated with a particular rule set, and as glossary entries [5].

These atomic and composite concepts are hierarchically arranged [25, 49], representing the application domain in a fine-grained DM. This allows the Web usage based adaptive systems to apply adaptation techniques to a low-level information structure, which leads to flexibility of the system and highly personalized solutions.

### 6.2.1.2   Relationships between the Domain Model Elements

Relationships are the tools used to generate meaningful new content in the DM by connecting two or more concepts together.

Relationships between elements are represented in the DM as the semantic relationships of the conceptualized information. Given this semantic component, meaningfulness depends on the application domain and Web Usage Based Adaptive Systems supporting goal [4, 11].

Amongst other configurations, relationships can be represented as objects that connect concepts by matching concept attributes, or as dependency graphs between groups of concepts [25, 49].

## 6.2.2   User Model

The main objective of a Web Usage Based Adaptive Systems is to deploy information in a personalized fashion, and this implies that the user is the central part of
such systems. By understanding the user's intentions and preferences, a Web Usage
Based Adaptive system can specify its own goals. This is why the User Model (UM)
is one of the main characteristics of these systems, since it allows the user to be vis-
ible for the system [25].

The UM describes the user from the system's point of view. This representation
is built by gathering user data resulting from the interaction between the system and
users. This data may be directly or indirectly generated by the users.

The UM consider users as data sets, typically stored as an attribute-value pair in
a table structure. These data can be divided into different types of data:

- User Data: This refers to user characteristics, which make every user a unique
  user. This data can be non-variant in time or may change over time. For instance,
  non-variant data are user profile data, while preferences may be considered as
  variant data [4].
- Usage data: Such data encapsulates the data produced by the interaction between
  users and the system. This is not necessarily data directly produced by the user,
  but it shows how the user utilizes the system and perceives the given content.
  Major examples of such data are the navigation behavioral data and the user
  browsing behavioral data [19, 40].

  – Navigation behavioral data: This is the information on how users move from
    page to page. This information can be obtained from the user logs on the web
    server.
  – User browsing behavioral data: This is the information on how users explore
    the content of a web page. This information can be obtained by means of user
    logs on the server or eye tracking devices, for example.

- User context data: This comprises all data aspects of the user's surroundings that
  are not included as User or Usage data. These define the context in which the
  user performs activities.

  Context has been defined as "...the set of the external parameters that can
  influence the behavior of the application by defining new views on its data and
  its available services. This parameters may be dynamic and may change during
  execution" [7]. Considering this definition, user context is defined by the user's
  physical surroundings, and available social networks and existing web services
  are part of the user context.

  User context data may be gathered by use of environmental sensors, different
  applications and other users surrounding the target user [30].

## 6.2.3   Adaptation Model

The way in which adaptation can be achieved depends on several variables, including the nature of the components that are available for adaptation and the goals that need to be completed after this process.

The adaptation model takes into account the above restrictions, processes them, and decides a feasible solution in order to deliver the rules that will lead to the adaptation.

In the next section, we present an overview of the main approaches in which the adaptation model has been applied.

## 6.3   Existing Approaches for Web Usage Based Systems

As web-based data is multidimensional, a huge number of possibilities appear when designing an adaptation strategy. Thus, a classification is useful in order to aid visualization and analysis of the different techniques in a heterogeneous space and for guidance in the decision process.

We based our classification on the taxonomy proposed by Knutov [26], which represents an evolution from the first ordering effort developed by Brusilovsky [8] in the sense of incorporating the concept of adaptive presentation along with content adaptation and navigation adaptation.

## 6.3.1   Content Adaptation

Content adaptation is defined as the process through which the information delivered by the web interface can be modified in order to fulfill an access requirement.

One of the main applications of content adaptation deals with the personalization of the content delivered to the user based on his usage. This idea is formally defined as the customization of a web site taking into account the knowledge extracted from the analysis of the user behavior [13].

There are technical challenges to delivering adequate usage-based strategies for content adaptation.

First, the strategies need to generate an artifact that represents the user need, in the form of a preference structure that serves as the main input to the core decision process of adaptation.

In that sense, usage data, which in web-based environments is commonly represented as a user session (the sequential path that the user follows during a visit to the website) has to be transformed efficiently in a vectorial representation that helps to identify which factors led the user to choose a particular set of pages when visiting a website.

Common approaches combine primarily the usage data and web page content to obtain the set of relevant keywords based on a frequency analysis using metrics such as TF-IDF. Other approaches use a topic-based analysis in order to reduce the dimensionality of the data generated.

Second, having settled a methodology to represent user preferences, a relationship between these preferences and the content must be established and the method for conducting the selection of content must be executed. Current approaches can be classified as follows:

### 6.3.1.1   Ontology-Based Selection of Content

Some ontology-based approaches include the use of user behavior, in the form of a model of the patterns followed by the user request, to generate new connections between the available data and the preferences. In [6], the authors propose the use of an iterative technique for monitoring the user navigation and to store it as preferences in the user's profile ontology . Then, they query this metadata in order to deliver personalized content.

### 6.3.1.2   Heuristic-Based Selection of Content

Heuristic techniques have gained a considerable amount of attention due to their high level of flexibility and low cost of execution. Those elements, and the convincing evidence of high level solutions despite the apparent simplicity, has led to their consideration as a feasible way to build content adaptation methods.

These methods work in a standard way, focusing on an exploratory phase in which some basic rules are learned in order to guide the behavior of the model. Subsequently, adaptation parameters are iteratively modified based on the feedback received from the user.

One of the first approaches in this area, called *AdPalette*, was proposed by [22], and consists of the use of Genetic Algorithms to customize the content of on-line advertisements based on web usage.

In *AdPalette*, each advertisement is modeled as a set of combinations of attributes (previously defined) and their values. Thus the main goal is to find the optimal configuration for each user preference. It is clear that this scenario configures a combinatorial optimization problem.

The approach proposed by the authors in [22], consists of using a genetic-based strategy for iteratively constructing the ads. The attributes are modified *crossover* and *mutation* operations from the standard implementation of an evolutionary algorithm, in order to adapt the content to maximize the response rates.

The technique can be summarized in the following steps, which were taken from [22]:

- Initialization: An initial set of $N$ ads is generated randomly. The response from the users is stored and rated based on a click through approach.
- Goal Program: Setting the attribute restriction to choose the set of best ads based on the responses from the users.

- Best ads selection: Based on the previous step, rank the set of ads and choose the best.
- Crossover: Switch the attribute levels between pairs of ads in order to generate new individuals with conceivably better attribute configuration.
- Mutation: Arbitrarily modify the values of a sub set of attributes in order to increase heterogeneity, which is useful for the exploration of new solutions.
- Ad placement: Set the ads in the web site and record the responses from the users.
- Deletion of poor individuals: The ads that do not achieve a minimum response must the discarded.

The experimental results showed that *AdPalette* represents an important improvement when compared with other techniques, including qualitative ones. In addition, an interesting discussion is established regarding the performance of the algorithm when using different combinations of the genetic operators.

Another bio-inspired approach is presented in [44], where White et al. proposed a novel technique based on Ant Colony Optimization for selecting on-line advertisements for a given user, trying to maximize the probability of clicking.

In this case, the model simulates the interaction between web users and the web server, which manages the page requests and decides which advertisements to match with the content that a given user is viewing.

Each user is modeled as a keyword-preference vector seeded from a pool of finite keywords. Likewise, web pages are also represented as vectors, and although in the paper the values are assigned randomly, the authors suggested that a Vector Space Model approach could be more reliable. In this way, applying user preferences, page content and advertisement content as inputs, the algorithm tracks the user behavior and iteratively generates matches between ads and content.

Although this approach seems promising, it was only evaluated and compared against a random selection of advertisements, and thus there is still some work that can be conducted in order to guarantee reliability.

Another relevant issue in which content adaptation represents a valuable solution is related with reducing web server overload. When the request rate on a web server increases beyond a manageable level, the server becomes overloaded and the quality of the service suffers intermittence, sometimes reaching a status of unresponsiveness. This leads clients to experience service outages, which could also represent a financial loss [1].

While the standard behavior of a server implies that most of the time the facilities remain unused, some particular events can produce a sudden rise of the number of requests. Such events vary from real world events, like catastrophes, in which case people suddenly demand information, to social media related phenomena, like the Slashdot effect, in which a standard website is promoted by a famous one, leading to high traffic in a short period of time [14].

In that sense, the most common approaches to reduce server load are related with the use of hardware-based solutions like caching and replication [36]. Content adaptation comes as a complementary alternative in order to contribute to a more dynamic and cost effective strategy [17].

136

In terms of the layer in which the adaptation is achieved, [21] classifies them as client-based, proxy-based and server-based, concluding that server-based strategies offer more advantages due to the fact that the server has access to more relevant data in order to deliver a web page with the optimal configuration, given dynamic restrictions.

In [18], a novel methodology is presented for dynamic delivery of webpages based on monitoring usage load. In the first place, the original webpage is uploaded to the server, and there several copies are generated. These new versions differ in terms of their quantity and quality of content; for instance, the quality of a photo is lowered sequentially using a standard filtering technique in order to decrease the total size of the load when the web page is requested. Then, the server is monitored and performance indicators are calculated. This quantitative measure deals mainly with the level of CPU utilization, the number of incoming TCP connections in a given unit of time, and the status of the existing running processes of the server.

Those metrics are used in a decision process that determines if switching between versions is necessary. The handling of the change between versions is performed using existing redirection modules like MOD REWRITE. Results showed that the optimized site could handle 2 to 4 times higher load than the normal configuration, using the same hardware.

## 6.3.2   Navigation Adaptation

Navigation adaptation handles the way in which the link structure of the website is modified in order to facilitate the search for information. The main idea is to develop a system that allows the extraction of user needs and to then use them to dynamically generate shorter paths.

### 6.3.2.1   Ontology-Based Techniques for Navigation Adaptation

Ontology-based techniques have also been implemented in order to optimize the navigation and sequencing of elements, specifically in the e-learning and tutoring system field [35].

The use of an ontology-based approach appears as a natural step after the proposal of rule-based techniques for adaptiveness in e-learning systems. In that sense, the work by Popescu et al. [32] represents one of the first attempts to formalize a framework to manage adaptiveness in educational environments. As can be seen in Figure 6.2, the main functionalities of the proposed framework are as follows:

• Tracking of user activity by monitoring the interaction with the system
• Identification of learning style based on usage
• Evaluation and implementation of the adaptation

Although the use of ontologies was primary focused on the development of frameworks that allowed the representation of educational materials and the visualization

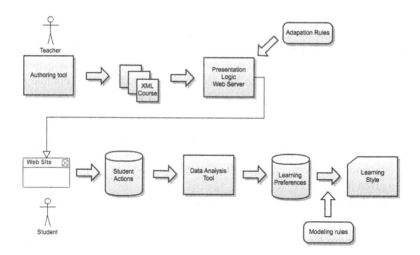

**Fig. 6.2** Rule-based framework for adaptiveness in learning environments

of their relationships, novel approaches have incorporated learner behavior in order to deliver a more effective user experience.

In [41], Vesin et al. proposed a technique based on the discovery of behavioral patterns and their use as the input to generate semantic rules through SWRL[1]. Then, the result is used to derive information that cannot be inferred with the basic description logic encapsulated within the ontologies. This rule-based reasoning is used to modify the link structure in order to deliver the pedagogical material in the optimal sequence that maximizes the learning process.

The core functionalities are executed in a sequence that proceeds from the recognition of the cognitive patterns of the users, through the identification of the domain and structural organization of the material and finally to the dynamic selection of the best sequence of learning content for each learning stage.

The key aspect of this adaptation process is dominated by three sets of rules:

- Learner-system interaction rules: Collect the usage data and build learner models.
- Offline rules: Use the previously created learned models to recognize users' goals based on their navigational sequences.
- Recommendation rules: Using the content that the user is interested in and the sequences tracked by the system, these rules provide a set of personalized learning objects.

### 6.3.2.2    Heuristic-Based Techniques for Navigation Adaptation

As hyperlink reconstruction can be analyzed as an optimization problem, several heuristic approaches have been proposed in order to tackle scalability issues related with the high level of data from hyperlink structure sources.

---

[1] Semantic Web Rule Language.

In [34], Saremi et al. proposed to model the improvement of a website as a Quadratic Assignment Problem (QAP). The key element in the representation of the problem is setting the goal that needs to be achieved. Qualitatively, the authors assumed that the position of the pages in the website impacts the navigational process, and thus, it is important to link closely-related webpages. From that starting point, a standard QAP formulation is adapted into a web structure context:

- $d(a_i, a_j)$, formally the distance between two facilities. In this case, it represents the number of required steps (links) to reach from page $i$ to page $j$.
- $C_{ij}$, the interaction cost between facilities. In this case, it represents the degree of connectivity between pages. It is calculated based on the probability of using a given link from $i$ to reach $j$.
- $TC(a)$, the total cost of the facility layout. In this case, the total cost of the website structure.

Given the above formulation, the objective function to be solved is repre- sented as below.

$$Min\left(TC(a) = \sum_{1 \leq i < j \leq n} W_{ij} \times d(a_i, a_j)\right) \quad (6.1)$$

A standard search-based technique is used to find an approximate solution. The results showed that the new structure reduced overall cost by almost 25% .

In [27], Lin proposed binary programming models for web graph reorganization, optimizing the cohesion of webpages based on real usage. Those models also incorporate the ability of reducing the following graph properties when performing the optimization process:

- The number of hyperlinks out of a particular node, because this means that the final user will be exposed to a page with too many links at the same time, which represents an information overload, which could undermine the decision process involved in the website navigation.
- The length of the shortest path between the home page to each node, which represents the number of links the user needs to follow in order to reach a given page.

The first model represents the problem of finding a subgraph that contains all the nodes and maximizes the sum of the link frequencies under defined degree and depth constraints.

$$Max \sum_{i,j \in E} f_{ij} x_{ij}$$

subject to

$$z_{ij} \leq x_{ij}, \forall (i, j) \in E,$$

$$\sum_{v_i \in B(v_j)} z_{ij} = 1, j = 2, ...n$$

$$\lambda_j m(1 - z_{ij}) \geq \lambda_i + 1, \forall(i, j) \in E,$$
$$\lambda_i \leq l, \forall v_i \in V,$$
$$x_{ij}, z_{ij} = 0, 1; \lambda_i \geq 0,$$

where $x_{ij}$ is a binary variable that represents the availability of a link between page $i$ and $j$. $d_i$ denotes the maximum allowable outward degree for $v_i$. $z_{ij}$ is a binary variable that takes the value 1 if the link between $i$ and $j$ belongs to a spanning tree within the subgraph. A spanning tree from a graph $G$ is a tree that is formed by all the vertices and a subset of the edges. It can be seen as the minimum set of edges that connect all vertices. $\lambda_i$ represents the level of $v_i$, and thus the constraints guarantee the order and also force the depth to a maximum of $l$.

As the above model requires a high amount of computation time and re- sources, the author proposed the separation of the constraints into two stages, the first one focused on discovering the spanning tree, and the second to find the subgraph that conforms to the out degree and depth restrictions. Although this separation reduced the computation time, the performance rapidly deteriorated when increasing the number of nodes.

To overcome this issue, the same author in [28] proposed the use of an Ant Colony Optimization approach to solve the problem instead of the deterministic method previously used. Results based on simulation and in a real website showed the feasibility of using Ant Colony Optimization for the reorganization of large websites.

In [29], Lingras et al. proposed a methodology to reconfigure link structure based on the web sequence pattern analysis. First, they state the common trade-off when selecting the optimal number of links: if the number is too big, it may overwhelm the user, but if it is too small, the user will be required to traverse longer paths in order to find the information that is sought.

As a solution to this problem, it was proposed to use the notion of the *neighbourhood* of a web page $p$. More precisely, the primary neighborhood, $N_1(p)$, is the set of pages that can be reached using one click. The secondary neighborhood, $N_2(p)$, is the set of pages that can be reached using two or fewer clicks. Similarly, the tertiary neighborhood, $N_3(p)$, is the subset of pages that can be reached using three or less clicks. Given the above definitions, the authors defined the following goals:

- Restrict the size of $N_1(p)$
- For each web access, $(p, v)$, where $p$ is a page and $v$ is an edge, then $v \in N_3(3)$

Then, in order to incorporate the web usage, each link is associated with a weight whose value is related with the frequencies of transition. To solve the selection problem, the concept of a minimum spanning tree is used, but this time extended in order to reach the restrictions exposed above.

In [23], Kazienko proposed a methodology to verify the usefulness of the hyperlinks in a website. The notion of verification is based on the analysis of the association rules that can be extracted from web usage.

Thus, a *positive* association rule, $X \rightarrow Y$ means that if $X$ occurs, then $Y$ co-occurs. This can be extrapolated to a web usage scenario, in the sense that if page $X$

(or a subset of pages) is present in a given session , then the set of pages $Y$ will also be part of that session.

Similarly, a *negative* association rule, $X \to\sim Y$, can be interpreted as if a set of pages $X$ is part of a given session, then the set of pages $Y$ is not likely to appear in that session.

Also, as the number of atomic association rules between pages may be large, the author proposed an integration mechanism based on the average calculation of the confidence measure for both positive and negative association rules.

The other key element that is incorporated in the model is the notion of a sequential pattern, which is basically defined as a sequence of items frequently visited one after another.

Thus, the combination of both elements allows the definition of the concept of *Sequential pattern with Negative Conclusions*, $q \to\sim X$, which means if the user visits a sequence $q$ of pages in a website, then he is not likely to visit the set of $X$. Thereby, with the new knowledge that can be extracted from this metric, it is possible to evaluate the usefulness of the link structure of a website.

### 6.3.3   Presentation Adaptation

As defined in [10], presentation adaptation consist of determining the way in which the information displayed in a website should be organized. Although most of the techniques developed are not mainly usage-based and they are more focused on context information and user needs [48], one of the problems usage-based presentation adaptation can contribute to solve is related to how to handle a subset of users with some impairments. For instance, one of the key questions in this area is how to automatically give support to elder users, who usually present a wide range of problems, from sensory to cognitive disabilities.

Thus, how the usage of a web platform can be applied to infer the level of accessibility that needs to be delivered to assist and improve the user experience [20].

The concept of *assistive technology*, introduced by Sloan et al. [37], deals with the classification of the levels in which technology is applied to accommodate a given impairment. Those levels are constructed in relation to the proximity and integration to the web interface that is being used, starting from the standalone, which represents just basic accommodations to some impairments, to the browser level, which are in more direct contact with users, affecting, for instance, the ability to enhance keyboard interaction or modify display characteristics.

Although many of the techniques have been implemented, there are still limitations and barriers that slow the use of assistive technology. In the first place, economic issues threaten access to more complex strategies for assistance, due to the fact that high costs could not be covered because the percentage of target users is not economically relevant. The second issue deals with the learning curve associated with the use of an assistive technology in relation with the way in which the user needs to interact.

Regarding more technical issues, [37] remarks the lack of awareness in such systems when they have to recognize that the user impairment has reached a level at which some aid is necessary. Thus, the challenging part for adaptation in this case starts with an accurate detection of the person's accessibility needs, and then, how to effectively select and implement an adaptation process in order to improve the user experience.

Trewin et al.[39] presented a novel technique called *Steady clicks* that tends to help users with motor impairments. Based on the usage of this kind of users, the main issue to attack was the involuntary slip of the cursor when pressing the mouse. To solve this problem, this technique generates a strategy that freezes the cursor and eventually avoids involuntary clicks.

Similar was the approach presented in [46], called *Angle mouse* in which the pointing action is divided into two phases: the first one, more general, deals with the initial approach to the target; the second one, called the corrective phase, is related to more fine adjustments in situations that could generate more difficulties for users with neuromotor disabilities. *Angle mouse* is able to analyze the behavior in the above situations, in terms of tracking the angles formed by the mouse and changing the target size in order to aid the user. Thus, the presentation of the content is dynamically changed in order to help the users.

In [16], Gajos et al. presented a technique to develop abstract interfaces that takes into account vision and motor limitations. Subsequently, in [15] the author presents a more systematic view on the automatic generation of interfaces given user disability restrictions, but it takes a more broad approach than just considering some usage-based metrics represented by the abilities, by also considering the user's devices, tasks, and preferences. The system presented is called *Supple* and it is based on an optimization approach that selects an interface from a set of feasible candidates in order to minimize the movement time of the user.

In order to improve the theoretical framework from the above results, [47] proposed the concept of *ability-based design* which aims to move from the standard notion on how to assist users when interacting with interfaces, to a more proactive approach in which the design is guided in order fit the abilities of any user.

## 6.4   Privacy Issues in User Data Collection and Usage

Privacy issues in the IT world appear in the complexity of social and computer systems interactions. Complex as these are, to treat them deeply goes beyond the scope of this work. Nevertheless, it is important to start building toward better understanding the extent of technology's impact on future society, as well as to prepare the technical IT community for a near future, where legal issues will heavily impact what was once purely technical work.

Previous Adaptive Hypermedia Systems overviews [25, 49] have mainly focussed on the technical aspects of the systems, missing to discuss this topic, but its treatment is much needed due to the nature of the interaction between users and computer systems. Since on the web this interaction happens mainly by means of browsing software, the data generated is easily linkable to particular users and/or groups of users [40]. Therefore, there is a risk of revealing information that may not be in the users interest to share [12].

From a technical point of view, it is important to clearly state basic concepts over which privacy issues arise. These concepts are i) understanding the object of discussion from a technical point of view, ii) appreciating where legal issues intersect with the technical field, and finally, iii) knowing how the technical community should behave to safely keep working without affecting people's right to privacy.

## 6.4.1   Generation, Access and Usage of Data

In this section we address the object of discussion: user data. In order to analyze data ownership and whom to blame for its misuse, it is important to understand what sort of data is being generated on the Internet, who generates it, how it is generated, and how it is being used.

### 6.4.1.1   What Sort of Data Is Generated on the Internet

The increasing usage of new technologies, such as web 2.0 and Ubiquitous Computing, have impacted the data originating from the Internet. This data is not only generated directly by humans, but includes traces of transactions between machines and users. The literature classifies these data into three categories: Content, Structure and Usage data [40].

- Content: Available objects within a web page, namely videos, sounds, pictures, text, etc. This information represents the resources perceived by the user.
- Structure: Hypertext structure defining a website. This information represents how the information is linked to form the website.
- Usage: Transactional data logged into a web server. This information represents the interaction between the user and the browsing software.

Not all of this data concerns user privacy issues. An important concept often used in related literature [45] is the concept of Personal Identifiable Information (PII). This concept refers to any data directly linked to an individual or allowing in some manner a connection or identification of a particular user.

PII results are central to numerous regulations, since they are the logical extension of privacy into the virtual world [12]. Much discussion exists around this concept, since without clearly identifying what constitutes PII, the idea of privacy is vague, too.

Data newly available via the Internet is not easily classifiable as PII, since it is not directly linked to a specific individual, but can be used for discrimination based on browsing patterns. As an example, is an IP address PII? [45] For some, it is not, since the IP address only identifies a computer that can be used by several users. For others, it is, because it is possible to classify a user session with the IP address and to infer information about the user (for instance, age or gender based on his/her clicks on a webpage).

As will be shown below, focusing on this discussion is not necessarily leading to a better understanding of privacy on the Internet, as there are other important issues that will contribute to defining new standards for privacy on the Internet, such as the purposes for the utilization of data and how such data is manipulated.

### 6.4.1.2   Data Processing in Web Usage Based Adaptive Systems

The personalization process performed by a Web usage based adaptive systems is mainly done based on sessionization [40]. By applying various techniques to mine usage data, a user's session is reproduced, so it can be understood by the system and later aggregated to other user's sessions to clusterize and understand common behavioral patterns.

It is interesting to notice that during this process, it is not relevant to identify a specific user, but to recognize repeated navigational patterns. In this way, a future user's behavior can be predicted, and users may be categorized into types based on behavior. Therefore, a user can be treated (in a positive or negative way) according to his/her user type.

## 6.4.2   What Is Private Data?

The Internet is a decentralized and deregulated network where the opportunity for easily and inexpensively tracing data from users is real [12], even without his/her consent. In addition, regulations are hard to implement, given the international nature of the Internet, and there is a great risk for fraudulent activities involving misuse of user data.

There is no international agreement on what constitutes an Internet privacy violation, and neither is there an exact understanding of the extent of such violations [12]. Therefore, we have witnessed different efforts seeking privacy protection and privacy statements to ease consumer concerns. Unfortunately, these efforts have not been sufficient to ensure full legal protection of user privacy. This may be due to the legal approach applied.

The classical approach to privacy is based on individual rights and responsibilities over personal data. In this way, Warren and Brandeis [42] defined privacy as the right to be alone. Later, William Prosser stated that privacy implies

...a complex of four interests... tied together by the common name, but otherwise [with] nothing in common. These interests being 1) intrusion upon the individual's seclusion; 2) public disclosure of embarrassing private facts; 3) publicity in a false light; 4) appropriation of an individual's attribute (such as name or likeness) [33].

Although such definitions may have proved useful in past circumstances, present technology proposes new threats to privacy. In the Internet environment, anonymity does not ensure good use of personal data. Profiling systems are an example, for these do not require identification of a specific user to sort and serve the user based on navigational patterns.

Considering the above, the main problem seems to be the definition of private information itself. What prior to the Internet era was easily identifiable is now unclear. Thus, so also are the implications of private information collection on privacy issues less than clear.

As stated in the literature, the path to a solution appears to be linked with the process by which the information is processed. A process-purpose classification is proposed in [], adding new dimensionalities to the study of privacy issues. These dimensions are i) why is information related to me (not necessarily private data) being requested, and ii) how will this data be processed?

### 6.4.3   Privacy-Friendly Development Policy

Palliative measurements for protecting user privacy will depend on the purpose and techniques used. As stated in [], two types of personalization can be identified, and different guidelines should be considered from the service provider side.

- Projects based on Web Logs with the intention of improving the navigation experience within a website:

  1. Publication of privacy policy detailing purposes and extraction patterns.
  2. Anonymity should be ensured.

- Projects based on mining tools on the web with the intention of making more complex inferences about an individual's attributes:

  1. Publication of privacy policy detailing purposes and extraction patterns.
  2. PII should be submitted so the user may choose weather to keep or quit the service at any time.

## 6.5   Future Challenges

The World Wide Web (WWW) represents a global platform for sharing resources, for information storage, and for service publications. The WWW, new network technologies, and new embedded devices are bringing the Internet out of cyberspace and into the physical world [43].

This trend represents a new boundary for Web Usage Based Adaptive Systems. As the Internet expanded Adaptive Hypermedia systems in the mid-1990's, new technologies as well as new interdisciplinary research have been expanding the Web Usage Based Adaptive Systems domain to the physical and social world.

Amongst other technologies, the Internet of Things, the Internet of Services, and Brain Informatics are introducing new questions and application dimensions for Web Usage Based Adaptive Systems. Some of the challenging issues that have arisen due to these new technologies are presented in this section.

- Internet of things (IoT): New wireless networks and tracking technologies (such as RFID [24]) allow the existence of a network of connected objects (things), such as appliances, mobile devices, and cars, etc. The IoT refers to the network composed by these uniquely identifiable objects and their virtual representations [30, 38].

  For Web Usage Based Adaptive Systems, the IoT represents a new source of resources, allowing the system to help the user alter the physical world by means of these objects in the IoT.

  Amongst others, important issues are i) how to include the connectivity, computation and data storage of the IoT into the Domain Model, ii) how to include the physical characteristics of the IoT into the Context Model and, iii) how to include quality attributes of these objects into the Adaptation Model to fulfill user preferences expressed in the User Model.

- Internet of services (IoS): A web service (WS) is a self contained piece of software that is remotely contacted and that enables communication between machines in a network. Using such WSs, multiple pieces of software can be orchestrated into a unique functionality that is coherent with the needs of the user [38].

  The concept of the IoS refers to the idea of an Internet where people look for services, instead of information based on a web page format. This concept reflects the current trend of apps utilization in mobile devices, where browsing is becoming less common. As mobile devices and embedded systems grow in popularity, the WWW will be used as a source of solutions instead of merely as a source of information, and the IoS will be dominant [30, 38].

  The IoS allows Web Usage Based Adaptive Systems to proactively interact with the user, since it allows the system to organize complete functionalities by only asking the final goal of the user. Since it is not necessary for the user to make decisions in the middle steps (such as choosing services), machine-user interactions are less stressful and more intuitive.

  Amongst others, important issues are i) how to incorporate WS functional attributes as part of the Domain Model, ii) how to include the WS composition process into the Adaptation Model, and iii) how to include the preferences of the user in the WS selection.

- Brain informatics (BI): This emerging field of study aims to understand the human information process mechanism by using experimental, theoretical, cognitive neuroscience, data mining, and machine learning techniques [30]. BI has the potential to increase the current understanding of the user, and therefore major changes to the User Model should be made.

Amongst others, important issues are i) how to include data from new technologies for tracking human behavior into the User Model, and ii) how to apply models from multiple fields of study for extracting computable information.

**Acknowledgements.** This research was supported by WCU (World Class University) program under the National Research Foundation of Korea and funded by the Ministry of Education, Science and Technology of Korea (Project No: R31-30007).

# References

1. Abdelzaher, T.F., Bhatti, N.: Web content adaptation to improve server overload behavior. Computer Networks 31(1116), 1563–1577 (1999)
2. Anagnostopoulos, I., Bielikova, M., Wallace, M., Lee, J.W.T.: Intelligent hypermedia for the adaptive web: Foreword to the smap '08 special session. In: International Workshop on Semantic Media Adaptation and Personalization, pp. 155–156 (2008)
3. Billsus, D., Brunk, C.A., Evans, C., Gladish, B., Pazzani, M.: Adaptive interfaces for ubiquitous web access. Communications of the ACM 45(5), 34–38 (2002)
4. de Bra, P., Houben, G.J., Wu, H.: Aham: a dexter-based reference model for adaptive hypermedia, pp. 147–156 (1999)
5. de Bra, P., Santic, T.: Aha! meets interbook, and more...
6. Brambilla, M., Tziviskou, C.: Modeling ontology-driven personalization of web contents. In: Eighth International Conference on Web Engineering, ICWE 2008, pp. 247–260 (July 2008)
7. Brusilovsky, P.: Methods and techniques of adaptive hypermedia. User Modeling and User Adapted Interaction 6(2-3), 87–129 (1996)
8. Brusilovsky, P.: Methods and Techniques of Adaptive Hypermedia. User Modeling and User-Adapted Interaction 6(2-3), 87–129 (1996)
9. Brusilovsky, P., Maybury, M.T.: From adaptive hypermedia to the adaptive web. Communications of the ACM 45(5), 21–24 (2002)
10. Bunt, A., Carenini, G., Conati, C.: Adaptive Content Presentation for the Web. In: Brusilovsky, P., Kobsa, A., Nejdl, W. (eds.) Adaptive Web 2007. LNCS, vol. 4321, pp. 409–432. Springer, Heidelberg (2007)
11. Cheverst, K., Mitchell, K., Davies, N.: The role of adaptive hypermedia in a context-aware tourist guide. Communications of the ACM 45(5), 47–51 (2002)
12. Chung, W., Paynter, J.: Privacy issues on the internet. In: Proceedings of the 35th Annual Hawaii International Conference on HICSS, p. 9 (January 2002)
13. Eirinaki, M., Vazirgiannis, M.: Web mining for web personalization. ACM Trans. Internet Technol. 3, 1–27 (2003)
14. Flake, G.W., Pennock, D.M.: Self-organization, self-regulation, and self-similarity on the fractal web. In: Lesmoir-Gordon, N. (ed.) The Colours of Infinity, pp. 88–118. Springer London (2010)
15. Gajos, K.Z., Weld, D.S., Wobbrock, J.O.: Automatically generating personalized user interfaces with supple. Artificial Intelligence 174(1213), 910–950 (2010)
16. Gajos, K.Z., Wobbrock, J.O., Weld, D.S.: Automatically generating user interfaces adapted to users' motor and vision capabilities. In: Proceedings of the 20th Annual ACM Symposium on User Interface Software and Technology, UIST 2007, pp. 231–240. ACM, New York (2007)

17. Gopshtein, M., Feitelson, D.G.: Empirical quantification of opportunities for content adaptation in web servers. In: Proceedings of the 3rd Annual Haifa Experimental Systems Conference, SYSTOR 2010, pp. 5:1–5:11. ACM, New York (2010)
18. Gopshtein, M., Feitelson, D.G.: Trading off quality for throughput using content adaptation in web servers. In: Proceedings of the 4th Annual International Conference on Systems and Storage, SYSTOR 2011, pp. 6:1–6:14. ACM, New York (2011)
19. Granka, L.A., Joachims, T., Gay, G.: Eye-tracking analysis of user behavior in www search. In: Proceedings of the 27th Annual International ACM SIGIR Conference on Research and Development in Information Retrieval, SIGIR 2004, pp. 478–479. ACM, New York (2004)
20. Hanson, V.L.: Age and web access: the next generation. In: Proceedings of the 2009 International Cross-Disciplinary Conference on Web Accessibililty (W4A), W4A 2009, pp. 7–15. ACM, New York (2009)
21. Jan, R.-H., Lin, C.-P., Chern, M.-S.: An optimization model for web content adaptation. Comput. Netw. 50, 953–965 (2006)
22. Karuga, G.G., Khraban, A.M., Nair, S.K., Rice, D.O.: Adpalette: an algorithm for customizing online advertisements on the fly. Decision Support Systems 32(2), 85–106 (2001); Decision Support Issues in Customer Relationship Management and Interactive Marketing for E-Commerce
23. Kazienko, P.: Usage-based positive and negative verification of user interface structure. In: Fourth International Conference on Autonomic and Autonomous Systems, ICAS 2008, pp. 1–6 (March 2008)
24. Khoo, B.: Rfid as an enabler of the internet of things: Issues of security and privacy. In: Internet of Things (iThings/CPSCom), 2011 International Conference on and 4th International Conference on Cyber, Physical and Social Computing, pp. 709 –712 (October 2011)
25. Knutov, E., De Bra, P., Pechenizkiy, M.: Ah 12 years later: a comprehensive survey of adaptive hypermedia methods and techniques. New Review of Hypermedia and Multimedia 15(1), 5–38 (2009)
26. Knutov, E., De Bra, P., Pechenizkiy, M.: AH 12 years later: a comprehensive survey of adaptive hypermedia methods and techniques. New Review of Hypermedia and Multimedia 15(1), 5–38 (2009)
27. Lin, C.-C.: Optimal web site reorganization considering information overload and search depth. European Journal of Operational Research 173(3), 839–848 (2006)
28. Lin, C.-C., Tseng, L.-C.: Website reorganization using an ant colony system. Expert Systems with Applications 37(12), 7598–7605 (2010)
29. Lingras, P., Lingras, R.: Adaptive hyperlinks using page access sequences and minimum spanning trees. In: IEEE International on Fuzzy Systems Conference of FUZZ-IEEE 2007, pp. 1–6 (July 2007)
30. Liu, J., Zhong, N., Yao, Y., Ras, Z.W.: The wisdom web: New challenges for web intelligence (wi). J. Intell. Inf. Syst. 20(1), 5–9 (2003)
31. Nejdl, W., Wolpers, M.: Kbs hyperbook - a data-driven information system on the web. In: 8th International World Wide Web Conference (1998)
32. Popescu, E., Badica, C., Trigano, P.: Rules for learner modeling and adaptation provisioning in an educational hypermedia system. In: International Symposium on Symbolic and Numeric Algorithms for Scientific Computing, SYNASC, pp. 492–499 (September 2007)
33. Prosser, W.: Privacy. Cal. Law Review 48, 383–389 (1960)
34. Saremi, H.Q., Abedin, B., Kermani, A.M.: Website structure improvement: Quadratic assignment problem approach and ant colony meta-heuristic technique. Applied Mathematics and Computation 195(1), 285–298 (2008)

35. Sicilia, M.-A., Lytras, M.D., Snchez-Alonso, S., Barriocanal, E.G., Zapata-Ros, M.: Modeling instructional-design theories with ontologies: Using methods to check, generate and search learning designs. Computers in Human Behavior 27(4), 1389–1398 (2011)
36. Sivasubramanian, S., Pierre, G., Van Steen, M., Alonso, G.: Analysis of caching and replication strategies for web applications. IEEE Internet Computing 11(1), 60–66 (2007)
37. Sloan, D., Atkinson, M.T., Machin, C., Li, Y.: The potential of adaptive interfaces as an accessibility aid for older web users. In: Proceedings of the 2010 International Cross Disciplinary Conference on Web Accessibility (W4A), W4A 2010, p. 35:1–35:10. ACM, New York (2010)
38. Soukkarieh, B., Sedes, F.: Towards an adaptive web information system based on web services. In: International Conference on Autonomic and Autonomous Systems, pp. 272–277 (2008)
39. Trewin, S., Keates, S., Moffatt, K.: Developing steady clicks: a method of cursor assistance for people with motor impairments. In: Proceedings of the 8th International ACM SIGACCESS Conference on Computers and Accessibility, Assets 2006, pp. 26–33. ACM, New York (2006)
40. Velásquez, J.D., Palade, V.: Adaptive web sitesa knowledge extraction from web data approach. In: Proceedings of the 2008 Conference on Adaptive Web Sites: A Knowledge Extraction from Web Data Approach, pp. 1–272. IOS Press, Amsterdam (2008)
41. Vesin, B., Ivanovic, M., Klasnja-Milicevic, A., Budimac, Z.: Rule-based reasoning for altering pattern navigation in programming tutoring system. In: 2011 15th International Conference on System Theory, Control, and Computing (ICSTCC), pp. 1–6 (October 2011)
42. Warren, S., Brandeis, L.: The right to privacy. Harvard Law Review 193(1) (1890)
43. Weiser, M.: The computer for the 21st century. IEEE Pervasive Computing 99(1), 19–25 (2002)
44. White, T., Salehi-Abari, A., Box, B.: On How Ants Put Advertisements on the Web. In: García-Pedrajas, N., Herrera, F., Fyfe, C., Benítez, J.M., Ali, M. (eds.) IEA/AIE 2010, Part II. LNCS, vol. 6097, pp. 494–503. Springer, Heidelberg (2010)
45. Whitten, A.: Are ip addresses personal? (February 2008)
46. Wobbrock, J.O., Fogarty, J., Liu, S.-Y(S.), Kimuro, S., Harada, S.: The angle mouse: target-agnostic dynamic gain adjustment based on angular deviation. In: Proceedings of the 27th International Conference on Human Factors in Computing Systems, CHI 2009, pp. 1401–1410. ACM, New York (2009)
47. Wobbrock, J.O., Kane, S.K., Gajos, K.Z., Harada, S., Froehlich, J.: Ability-based design: Concept, principles and examples. ACM Trans. Access. Comput. 3, 9:1–9:27 (April 2011)
48. Zakraoui, J., Zagler, W.: A Logical Approach to Web User Interface Adaptation. In: Holzinger, A., Simonic, K.-M. (eds.) USAB 2011. LNCS, vol. 7058, pp. 645–656. Springer, Heidelberg (2011)
49. Zemirline, N., Bourda, Y., Reynaud, C.: Leveraging Adaptive Web with Adaptation Patterns. Technical report 1529 (November 2009)

# Chapter 7
# Recommender Systems: Sources of Knowledge and Evaluation Metrics

Denis Parra and Shaghayegh Sahebi

**Abstract.** *Recommender* or *Recommendation Systems* (*RS*) aim to help users dealing with information overload: finding relevant items in a vast space of resources. Research on *RS* has been active since the development of the first recommender system in the early 1990s, Tapestry, and some articles and books that survey algorithms and application domains have been published recently. However, these surveys have not extensively covered the different types of information used in *RS* (sources of knowledge), and only a few of them have reviewed the different ways to assess the quality and performance of *RS*. In order to bridge this gap, in this chapter we present a classification of recommender systems, and then we focus on presenting the main sources of knowledge and evaluation metrics that have been described in the research literature.

## 7.1 Introduction

*Recommender* or *Recommendation Systems* (*RS*) aim to help a user or a group of users in a system to select items from a crowded item or information space [70]. In order to generate recommendations, a *RS* might try to match users' characteristics with items' characteristics by performing content filtering, or it might look at previous interactions of the user in the system to match users with similar patterns [53]. A typical domain where RS are useful is the World Wide Web (WWW): with its

Denis Parra
School of Information Sciences, University of Pittsburgh
135 North Bellefield Avenue, Pittsburgh, PA 15260
e-mail: dap89@pitt.edu

Shaghayegh Sahebi
Intelligent Systems Program, University of Pittsburgh
Sennott Square, Pittsburgh, PA 15260, USA
e-mail: ssahebi@cs.pitt.edu

J.D. Velásquez et al. (Eds.): Advanced Techniques in Web Intelligence-2, SCI 452, pp. 149–175.
springerlink.com                                      © Springer-Verlag Berlin Heidelberg 2013

overwhelming growth of available information and the continuously growing number of different devices that can be used to access it *RS* have taken on an important role in the daily lives of people to find relevant resources, such as movies [41], books [56], music [18], tourism destinations [12], or cooking recipes [26].

The first recommender system, Tapestry [32], was introduced almost 20 years ago by Goldberg et al. to deal with the increasing amount of messages that users received by email. This early system –as well as GroupLens developed by Paul Resnick et al. [96] and Ringo by Shardanand and Maes [107]– made use of a technique called *Collaborative Filtering* (CF) to provide recommendations to a center user based on previous actions performed by herself and by like-minded users, denoted as nearest neighbors. All these systems make use of some form of deviance measure between a predicted and a real value of preference for evaluation. In their seminal paper, Herlocker et al. [42] survey different tasks and metrics for *RS*, introducing, among others, the concepts of serendipity and novelty. However, these concepts started to have a larger impact on the evaluation of *RS* after the Netflix prize.

The *Netflix Prize*[1] was a contest created by the movie rental company Netflix[2] in October of 2006 [11]. The *Netflix Prize* challenged the data mining, machine learning and computer science communities to improve the algorithm Cinematch by at least 10% in terms of predicting the ratings that users assigned to movies. The winners of this challenge would receive a $1 million dollar prize. Netflix released a dataset of 100 million anonymous movie ratings and the evaluation was based on Root Mean Square Error (*RMSE*), a metric that we explain in section 7.4.1. Although the community of researchers engaged in *RS* existed well before this contest, the *Netflix Prize* attracted a large amount of people from different areas. It might not be a coincidence that the ACM Recommender Systems conference, targeted specifically for *RS*, began in 2007. Despite the benefit of attracting a large community of researchers to the field, the Netflix Prize had the negative effect of focusing on accuracy in the active evaluation period, giving less importance to important characteristics of the recommendations such as coverage, novelty, or diversity. By the time the challenge was finished, the *RS* community started to show more interest in other quality metrics.

Some studies have gone beyond accuracy to evaluate *RS* such as recommendation diversification by Ziegler et al. in 2005 [128] and Zhou et al. in 2010 [125], serendipity by Murakami et al. in 2008 [80] and by Zhang et al. in 2011 [124], and coverage by Ge et al. in 2010 [29]. More recently Vargas and Castells try to combine accuracy and serendipity in a single evaluation framework [113]. These new trends in *RS* evaluation stem from several factors, among which we count:

- **Accuracy and user satisfaction are not always related**: Some articles showed that rating prediction accuracy is not always correlated with other metrics [95], and most importantly, not necessarily correlated with user satisfaction [39] [70]. This result supported the need for creating new evaluation measures that better

---

[1] http://www.netflixprize.com

[2] http://www.netflix.com

predicted the final goal which is a user-centric evaluation of the *RS* rather than only an off-line evaluation.

- **Lack of explicit user feedback**: Although curiosity is a human trait, turning users from lurkers into real contributors to a system is a challenging task [92]. For this reason, algorithm and evaluation metrics that rely on implicit user feedback have become more frequent in recent years.
- **New sources of knowledge**: In the early days of *RS*, two contemporary popular technologies were not available: Smartphones and social networks. The first can provide a good deal of contextual information, such as temporal data, location, and additional ways to interact than a desktop computer does. The second, social networks, provides contextual information that impacts the development of trust-based methods: real family and friends. In addition, users contribute with long-term (birthday, preferred sports, art, or politics) and short-term information (*likes* on a specific comment or picture), giving *RS* different signals to produce recommendations.

In the following sections, we review *RS* by presenting a classification in section 7.2. Then, in section 7.3 we describe the main sources of knowledge used to provide recommendations, to continue with section 7.4 presenting the metrics used to evaluate quality and performance of *RS*. In section 7.5, we present all of the aforementioned concepts in the context of Web Recommendation, and we finalize summarizing the chapter adding a list of ongoing and future challenges in this area.

## 7.2    Classification of Recommender Systems

The ultimate goal of any user-adaptive system is to provide users with what they need without asking them explicitly [79] [115]. This identifies the difference between personalization and customization. The difference between these two is in the actor who controls the creation of user profiles as well as the presentation of interface elements to the user. In customization, the users usually control their preferences or requirements manually. On the other hand, in personalization, the user profiles are created and potentially updated by the system automatically and with minimal explicit control by the user [73]. These systems can reduce the amount of time a user spends to find her required items [27]. The process of web personalization is consisted of three phases: data preparation and transformation, pattern discovery, and recommendation [81]. In traditional collaborative filtering approaches, the pattern discovery phase (e.g., neighborhood formation in the k-nearest neighbor method) as well as the recommendation phase is performed in real time. In contrast, personalization systems which are based on web usage mining, perform the pattern discovery phase in an online state. The data preparation phase transforms raw web log files into clickstream data that can be processed through data mining tasks. A variety of data mining techniques can be applied to the clickstream or Web application data in the pattern discovery phase, such as clustering, association rule mining, and sequential pattern discovery. A recommendation engine considers the active user session in conjunction with the discovered patterns to provide personalized content [116]. The personalized content can take the form of recommended links or products, or targeted advertisements [81]. At first, traditional Recommender Systems were defined as systems that collected user opinions about various subjects and guided users towards their items of interest. This was done using collaborative filtering approaches [96], [97]. After a while, these systems started using broader research approaches and played a more active role related to users. As a result, any system that produces individualized recommendations as its output or has the effect of guiding users to interesting or useful objects is defined as a personalization system [16]. Generally, personalization is based on a mapping between users and items to interest values [3]. The learning process of Recommender Systems is divided into two general methods: memory-based (lazy-learning) Recommender Systems and model-based Recommender Systems [73]. In memory-based models, the entire data is stored and used in the memory while calculating the recommendations. As a result, these systems are sensitive to scalability issues. On the other hand, the expensive learning process in these systems gets completed offline. Model-based systems are more scalable in high data volumes.

Generally, recommender systems are divided into three groups based on their input data type, approaches to create user profiles, and algorithmic methods utilized to produce recommendations: rule-bases, content-based, and usage-based systems [73]. Each of these three groups are discussed in the following sections.

## 7.2.1  Rule-Based Recommender Systems

In rule-based recommender systems, decisions are made based on some rules that are extracted, either manually or automatically, from user profiles. The goal in these systems is to find factors that influence users' choice of an item or product. Many of the existing e-commerce websites use manual rule-based recommender systems. These systems permit the site administrators to set the rules based on statistical, psychological, and demographic information about users. In some cases, the rules are very domain dependent and reflect the business goals of the website. These rules are used to improve the contents provided to a user when her profile matches at least one of the conditions. Like many other rule-based systems, this method of recommendation depends on the knowledge engineering abilities of the system designers to build a suitable rule-base for specific characteristics of the domain and market. User profiles are usually achieved by explicit interaction with users. Some research has been done on the learning methods for categorizing users into different groups based on their statistical information and then inferring the required rules for recommendation [90]. These methods aim to extract personalized rules for each user by use of reasoning approaches [17]. The general mechanism in these systems is that the user announces her interests to the system and then the system assesses each of existing items for each user, based on the knowledge base it has. We can name ISCREEN [91] as one of the rule-based systems that uses manually generated rules to filter its messages. Another example is Expertise Recommender [69] which recommends expert software engineers to programmers, based on the problems they report in programming. One of the advantages of these systems is the users' capability to express characteristics of their favorite items. One of the problems in these systems, in addition to the limitations of knowledge engineering, is the method used to generate user profiles. The input to these systems is user explanations about their personal interests and as a result, it is a biased input. Profiles in these systems are usually static and consequently, the performance of the systems degraded is by time passing and aging user profiles.

## 7.2.2 Content-Based Recommender Systems

Content-based Recommender Systems provide recommendations to users based on comparing items or products to the items that user showed interest in. A user profile in these systems represents explanations of product characteristics that user chose before. These explanations are illustrated by a set of characteristics or features describing the products in a user profile. The act of producing recommendations usually includes comparing features of items unseen or unrated by the user with her profile's content description. The items that are similar enough to the user's profile are recommended to her.

Content-based recommender systems usually rely on Information Retrieval techniques such as classification, clustering and text analysis [77]. In most of the content-based recommender systems, especially in the web-based and e-commerce systems, content descriptions are textual features extracted from web pages or product descriptions. Typically these systems rely on known document modeling approaches, which are rooted in information retrieval and information filtering research [99] [10]. User profiles and items can be shown as weighted vectors of words (e.g. based on *tf.idf* weightening model). Predicting a user's interest in an specific item can be done based on calculating vector similarity (such as cosine similarity measure) between the user profile vector and the item profile vector or based on probabilistic methods (such as bayesian classifiers). Additionally, despite collaborative filtering methods, user profiles are created individually, based only on the items seen or rated by the user himself/herself.

We can name Letizia [65], NewsWeeder [57], Personal WebWatcher [71], InfoFinder [55], and Syskill-Webert [89] among the first examples of content-based recommender systems.

One of the problems in content-based recommender systems, due to relying on user's previous ratings and interests, is the tendency to specification in choosing items [72]. However, user studies show that users tend to be more interested in novel and surprising items suggested by recommender systems [108]. Additionally, the practical relationships between items, such as their co-occurrence of use, or being complements for accomplishing a specific task, is not considered here. Another problem is that some items based cannot be represented with specific features, such as textual, so they won't be available in these recommender systems.

## 7.2.3   Collaborative Filtering Recommender Systems

Collaborative filtering [41] aims to solve some of the problems in rule-based and content-based recommender systems. Collaborative filtering-based recommender systems have achieved an acceptable success in e-commerce sites [104]. These models usually include matching item ratings of the current user (like ratings on books, or movies) to similar users (close neighbors) to recommend items that are not yet seen/rated by this user. In the standard case, these systems are memory-based. Traditional collaborative filtering systems used a standard memory-based classification approach based on $k$-nearest neighbor ($k$NN) method. In this algorithm, the target user profile is compared to other user profiles to identify the first $k$ users who have similar interests to this user. In traditional collaborative filtering, the predicted rating of active user $a$ on each item $j$ is calculated as a weighted sum of similar users' rankings on the same item: Equation 7.1. Where $n$ is the number of similar users we would like to take into account, $\alpha$ is a normalizer, $v_{i,j}$ is the vote of user $i$ on item $j$, $\bar{v}_i$ is the average rating of user $i$ and $w(a, i)$ is the weight of this $n$ similar users.

$$p_{a,j} = \bar{v}_a + \alpha \sum_{i=1}^{n} w(a, i)(v_{i,j} - \bar{v}_i) \tag{7.1}$$

The value of $w(a, i)$ can be calculated in many ways. Common methods are Cosine similarity, Euclidean similarity, or Pearson Correlation on user profiles.

Although these systems aim to provide a solution to issues in previous models of recommender systems, they suffer from their own problems. The most important problem of traditional memory-based collaborative filtering systems is that they are not scalable. In the $k$NN algorithm, formation of neighbors should be done in an online method. In other words, contrary to the model-based methods in which the model learning phase is done offline on the training data, the modeling phase in these systems is performed as an online task. With increase in users and items, this method can be unacceptably slow to produce dynamic recommendations during the interaction with users.

Another problem is due to the sparse nature of most of the datasets. More items in the dataset result in a decreased density of the user profile. As a consequence, the probability of similarity of seen items among users decreases, which results in less confidence in correlation calculations. Besides, collaborative filtering models perform at their best when there are explicit non-binary ratings for items while it is not the case for many websites. In some websites collecting user information for personalization is easier to be done using visited pages or products or asking for a product's information or changes in the shopping cart. These sources of information are considered as implicit feedback, which is discussed in section 7.3.

This method also suffers from the "new item" problem. When a new item or product is added to the item-set, it has never been seen or rated by any users. As a result, it does not exist in any user profile and the recommender system cannot recommend it to any user. The lack of ability to explain recommendations to users is another problem of these systems. Since collaborative filtering recommender systems do not use other information resources, like the content or semantic data, they cannot explain the reason for recommending a specific item to user.

To solve the sparsity and scalability problems, some use optimization techniques [5] [103] [123]. These methods include dimensionality reduction techniques, similarity indexing, and offline clustering of user profile in the past to search in the matched cluster while generating recommendations.

Another method which is based on collaborative filtering is item-based collaborative filtering [102]. In this method, a similarity matrix of items is produced based on rating data of user profiles in an offline way. This matrix is used to generate recommendations in the online phase. In other words, instead of relying the similarity between items in their content descriptions, it is calculated based on user ratings of them. Each item is shown as a vector and the similarities are calculated based on measures such as cosine similarity or based on correlation-based similarities such as Pearson or Spearman correlation. The process of generating recommendations predicts the rating of the target user to an unseen target item, by a weighted sum of given ratings to similar items to the target item. The same can be done on the item profiles. Evaluation of this method shows that it can produce recommendations with similar qualities to the model-based collaborative filtering recommendations [19].

Most of the personalization data mining methods are an extension of collaborative filtering. In these methods a pattern recognition algorithm takes prior user profiles or ratings as its input and generates an aggregated model of users. These models can be used with the current user profile to generate recommendations or predict user behavior in the future.

### 7.2.4 Hybrid Recommender Systems

As mentioned in the past sections, both content-based and collaborative filtering recommender systems have their own problems. Content-based recommenders cannot capture and utilize various types of similarities such as co-occurrence among items. Collaborative filtering methods have the "new item" problem. Hybrid recommender systems aim to solve the problems of content-based and collaborative filtering recommenders by use of various sources of information and combining both methods [63] [20] [21] [76]. They use both usage data of users and content data of items. Consequently, in addition to capturing the content similarities between items, these systems are able to reveal other relationships, such as associations and co-occurrences, between them. Another new direction in hybrid recommender systems is in using semantic web mining to extract semantic relationships between users and items [14] [9] [126]. Since using only keywords in finding similarities between objects has problems such as polysemy and synonymy, these models use the domain knowledge in form of a dictionary, ontology, or concept hierarchy to solve them. Some of these systems have used other sources of information such as the hierarchical link structure of a website as an additional domain knowledge [82] [98]. In general, these systems showed better results in predicting user interests.

## 7.3   Sources of Knowledge for Recommender Systems

### 7.3.1   Ratings

Ratings have been the most popular source of knowledge for *RS* to represent users's preferences from the early 1990s [96], [107], [101], to more recent years [61], [2], [51], [54]. The foundational *RS* algorithm collaborative filtering, presented in section 7.2.3, tries to find like-minded users by correlating the ratings that users have provided in a system. The goal of the algorithm is predicting users' ratings, under the assumption that this is a good way to estimate the interest that a user will show for a previously unseen item. This *rating prediction* task was the main objective of the Netflix Prize, and new algorithms were created that significantly improved the performance of the *Cinematch* algorithm. However, it has recently been shown that relying on additional information about the user or her context improves the performance of RS [4], [28]. Furthermore, in numerous occasions there are no ratings available and methods based on implicit feedback must be used [44]. The following sections describe these additional or alternative sources of knowledge.

### 7.3.2   Implicit Feedback

This source of knowledge refers to actions that the user performs over items, but that cannot be directly interpreted as explicit interest, i. e., the user explicitly stating her preference or the relevance of an item. This characteristic may seem as too noisy to consider using it in recommendations, however, mapping implicit and explicit feedback has been studied for several years, showing a strong correlation between both that makes implicit feedback a suitable source of knowledge to represent users' interests. Already in 1994, Morita and Shinoda [78] proved that there was a correlation between reading time on online news and self-reported preference. Konstan et al. [49] did a similar experiment with the larger user base of the Grouplens project and again found this to be true. Oard and Kim [83] performed experiments using not only reading time, but also other actions like printing an article, to find a positive correlation between implicit feedback and ratings.

Lee et al. [60] implement a recommender system based on implicit feedback by constructing "pseudo-ratings" using temporal information. In this work, the authors introduce the idea that recent implicit feedback should contribute more positively towards inferring the rating. The authors also use the idea of distinguishing three temporal bins: old, middle, and recent. Two recent works approach the issue of implicit feedback in the music domain. Jawaheer et. al analyze the characteristics of user implicit and explicit feedback in the context of last.fm music service [47]. However, their results are not conclusive due to limitations in the dataset since they only used explicit feedback available in the last.fm profiles, which is limited to the love/ban binary categories. This data is very sparse and, as the authors report, almost non-existent for some users or artists. On the other hand, Kordumova et. al use a Bayesian approach to learn a classifier on multiple implicit feedback variables [50]. Using these features, the authors are able to classify liked and disliked items with an accuracy of 0.75, uncovering the potential of mapping implicit feedback directly to preferences. In the music domain, Parra et al. [85] [87] mapped implicit feedback to explicit preference on the consumption of music albums. They found a significant effect of the number of times people listened to music and how recently the did it on the users' explicit preference (users' ratings). In a different domain, Fang and Si [23] propose a matrix co-factorization method that integrates user profile information and implicit feedback to provide recommendations of articles in the scientific portal nanohub.org.

### 7.3.3 Social Tags

Social Tagging systems (STS) allow users to attach free keywords, also known as tags, to items that users share or items that are already available in the system. Common examples of these systems are CiteULike[3], Bibsonomy[4] , or Mendeley[5] (mainly for academic resources), Delicious[6] (URLs), Flickr[7] (photographs), and last.fm (music). In these systems, the primary user action is the "social annotation" or "instance of tagging", corresponding to a tuple $(u, i, t)$ where $u \in$ Users, $i \in$ Items, and $t \in$ Tags. These systems have been studied in IR (Information Retrieval) to assess their potential to improve web search. Although there are some limitations especially in terms of coverage, as social bookmarking systems capture a rather small portion of the World Wide Web, they have shown promising results [43] [120].

In these systems, the recommendation of tags and resources (urls, photographs, academic articles) has several years of research. In [46], Jschke et al. evaluate tag recommendations comparing simple heuristics methods with an adapted user-based CF method, and FolkRank, which became state-of-the-art algorithm for tag recommendations. Furthermore, Tso-Sutter et al. [112] go further by using the user

---

[3] www.citeulike.org

[4] www.bibsonomy.org

[5] www.mendeley.com

[6] www.delicious.com

[7] www.flickr.com

annotations to recommend items (flickr photographs) instead of tags. They evaluate several methods using recall, and the best performing one is a method that "fuses" *user* x *item*, *item* x *tag*, and *user* x *tag* dimensions. Bogers [13] performs several evaluations combining and comparing content-based information with usage-based approaches. He uses MAP (Mean Average Precision) as fundamental evaluation metric, finding positive results for methods that fuse content and usage information, but he also warns about the spam and duplicates in the social bookmarking systems as a major threat to its more wide usage as source of user interest. Parra and Brusilovsky [86] also propose two variations of user-based collaborative filtering (CF) by leveraging the users' tags in citeulike to recommend scientific articles, showing that the proposed tag-based enhancements to CF result in better precision, rank and larger coverage than traditional rating-based approaches when used on these collections.

### 7.3.4   Online Social Networks

Social Recommender Systems (SRSs) are recommender systems that target the social media domain [34]. The main goals for these systems are to improve recommendation quality and solve the social information overload problem. These recommender systems provide people, web pages, items, or groups as recommendations to users. They use familiarity [36] [38], as connections on social web, similarity of users who might not be familiar with each other [35] [62], and trust [59] [6] as useful features of the social web. Also, a combination of these different features can be used in a hybrid social recommender system [37].

Social recommender systems can be categorized by three groups: social recommenders for recommending items, social recommenders for recommending people, and group recommender systems. In the first category, social relationships help collaborative filtering approaches to find more accurate recommendations [31] [33]. These recommendations can come from people the user knows and thus can judge them easily. They are based on both familiarity and similarity factors and as a result they are more effective for new users. In [38], Guy et. al. showed that familiarity results in more accurate recommendations while similarity results in more diverse items.

Group Recommender Systems (GRSs) provide recommendations to a group of people. Polylens was an early group recommendation system evaluated on a large scale, built to recommend movies to groups of people [84]. In the study, O'Connor et al. showed that users value the system, and are even willing to yield some privacy to get the benefits of group recommendation. In [105] , Senot et al. evaluate different group profiling strategies on a large-scale dataset of TV viewings, showing that the utilitarian strategy was the best but acknowledging that further study was needed to generalize the results to other domains. Another study by Baltrunas et al. show that when individual recommendations are not effective, group recommendation can result in better suggestions [7].

*Trust.* An important line of research in *RS* has been the influence of trust in the decisions the user makes to choose recommended items. Goldbeck adopts Sztompka's definition of trust in a research where she performs several experiments relating trust, similarity and derivations of trust from either one: "Trust is a bet about the future contingent actions of others" [30]. The influence of trust and its relationship with similarity have been already shown by Sinha and Swearingen, where people tended to prefer recommendations from friends than from systems, suggesting that it is because people have more trust for friends. This connection was most strongly clarified by Ziegler and Goldbeck, showing that the more similar two people were, the greater the trust between them [127]. Similarity is one of the core components of Collaborative Filtering, but Goldbeck's results show that trust captures more nuanced facets of correlation between users in a system than only similarity [30]. Other important works in this area include Massa and Avesani's research showing how some weaknesses of *RS* can be effectively alleviated by incorporating trust [68], and also Walter et al. who investigates a model of trust-based *RS* with agents that use their social network to reach information and their trust relationships to filter it [118].

One of the main drawbacks of this technique, as pointed out by Victor et al. in [117], is the lack of publicly available datasets (other than Epinions.com, the most used on this area) that allow to test trust-based approaches.

## 7.3.5 Context

### 7.3.5.1 Location

Unlike years ago, location information about the users is now widespread with the proliferation of mobile devices that incorporate GPS technology. This has allowed the field of *RS* to incorporate this information in the recommendation process, either as the single input information or as a complementary source of knowledge. One of the earliest systems to consider location to provide recommendation in a mobile-device was CityVoyager [110] which recommended places to shop in Tokyo. The design of the system was innovative, but the user study was too small to generalize results. They asked 11 users to freely shop and evaluate their shopping experience –the shopping stores–, and with the data gathered they tuned a recommendation model and evaluated the recommendation with just two users.

Another location-aware shopping system was developed and evaluated by Yang et al. [121]. In this casev they proposed a system for recommending vendors' web-pages –including offers and promotions– to interested customers. They compared four recommendation approaches (content-distance-based, content-based, distance-based, and random) in a user study with 136 undergraduate and graduate students that used the system for a period of a year and a half (January 2004 to August 2005). The evaluation measured satisfaction of the recommendations, and the content-distance-based approach had the best results overall. A more recent work by Quercia et al. [94] studied the recommendation of social events in the Boston, MA area using a mobile location-aware recommendation system. They sampled the location estimation of one million mobile users, and then combined the sample with social events in the same area, in order to infer the social events attended by 2,519 residents. Upon this data, they tested a variety of algorithms for recommending social events and found that the most effective algorithm recommends events that were popular among residents of an area. The least effective, instead, recommends events that are geographically close to the area. They evaluated the quality of the recommendations through several variations of percentile-ranking, the same metric used by Hu et al. in [44] and Fang and Si in [23], but under a different name.

### 7.3.5.2 Time

Although time or temporal information cannot always be considered directly as a source of preference, several methods and systems make use of time in their recommendations, especially in combination with other sources of user interest. As already mentioned in the section 7.3.3 regarding implicit feedback, Lee et al. [60] conflate implicit feedback and temporal information in a mobile e-commerce site, measuring its success by the increase in sales per recommendations provided. Another successful method incorporating time is TimeSVD++, introduced by Koren in [52], which accounts for temporal effects in the rating behavior of users and the rating pattern for items over the time. In a different approach, Lathia et al. [58] present a study of temporal effects in user preference. They study the effect on recommendations given that users continue to rate items over time, and they also investigate "the extent that the same items are being recommended over and over again". In the article, they also introduce two metrics to measure diversity and novelty, which are described in the section 7.4.

### 7.3.6 Heterogeneous Sources of Knowledge

Combining different sources of information has proven to be beneficial in some research cases. Fernandez-Tobias et al. present a cross-domain approach based on information obtained from the Linked Data project [25]. Using semantic representations, the authors recommend music artists based on places of interest: music venues. Another interesting case of heterogeneous data usage is the one presented by Fazel-Zarandi et al., which provides personalized expert recommendation based on semantic-data, a theoretical framework of social drivers, and social network analysis which shows promising results [24].

## 7.4 Evaluation Metrics for Recommender Systems

Although accuracy metrics have been frequently used to evaluate *RS* [15, 96, 107, 40], there are more dimensions that need to be assessed to capture their performance.

In a broad sense, the paper written by Herlocker et al. in 2004 [42] is a cornerstone for the evaluation of *RS*, as it describes several recommendation tasks that go beyond providing a plain list of recommended items, and many more evaluation metrics than accuracy. From this paper and further research stem the idea that the quality of a *RS* as perceived by a user is related to additional characteristics such as diversity of the recommended items [128], or how much user information and feedback needs the *RS* to perform well [111]. In the upcoming subsections, we describe several measures that have been used to evaluate these dimensions. Moreover, we include in the Section 7.4.5 the description of two frameworks recently introduced that fill the gap in the evaluation of the user experience of *RS*.

## 7.4.1   Prediction-Based Metrics

Prediction metrics allow one to compare which *RS* algorithm makes fewer mistakes when inferring how a user will evaluate a proposed recommendation. Predicting the ratings that a user will give to an item is the main optimization performed in rating-based *CF* recommender systems. The first of these measures is the *Mean Absolute Error* (*MAE*), which measures the mean of the absolute deviance between the predicted and the actual rating given by the users in the system.

$$MAE = \frac{\sum_{i=1}^{N} |p_i - r_i|}{N} \qquad (7.2)$$

In equation 7.2, $p_i$ is the predicted rating, $r_i$ is the actual rating and $N$ is the total number of predictions. In order to give more importance to cases with larger deviances from the actual ratings, *Mean Squared Error* (*MSE*) is used instead of *MAE*.

$$MSE = \frac{\sum_{i=1}^{N} (p_i - r_i)^2}{N} \qquad (7.3)$$

A variant of *MSE* is the *Root Mean Squared Error* (*RMSE*), which was the error metric used in the Netflix Prize.

$$RMSE = \sqrt{MSE} \qquad (7.4)$$

## 7.4.2   Information Retrieval Related Metrics

In an scenario where a user is provided with a list of recommendations in which she can evaluate the items as relevant or not relevant, metrics used in information retrieval such as Precision, Recall, or DCG are useful to assess the quality of a recommendation method. For instance, tag-based recommendations rely heavily on these metrics since users do not usually state their preference by rating the items [13, 86].

*Precision* is the fraction of recommended items that are relevant [67]. It is defined as

$$Precision = \frac{|\text{relevant items recommended}|}{|\text{items in the list}|} \tag{7.5}$$

The number of items recommended in a list can be very high depending on the recommendation method and the size of the dataset, and it is not feasible that a user will be able to check and evaluate all of them. For that reason, the evaluation metric will consider only the top items, which is called Top-N recommendation [19], and it is usually presented in articles as *Precision@n*. Precision or precision@n are used to evaluate the system in the context of a single user. In order to obtain a single metric that accounts for the precision of the recommendation method over the whole set of users, *Mean Average Precision (MAP)* is used. *MAP* is obtained by calculating the mean over the average precision of the list of recommendations from each user, as

$$MAP = \sum_{n=1}^{N} \frac{AveP(n)}{N} \tag{7.6}$$

In the equation, *AveP(n)* is the average precision for user *n*, i.e., the average of the precision values obtained for the set of top-N recommendations after each relevant recommendation is retrieved [67].

*Recall* is another typical metric used in information retrieval. It is defined as the fraction of relevant recommendations that are presented to the user [67]

$$Recall = \frac{|\text{relevant items recommended}|}{|\text{relevant items}|} \tag{7.7}$$

However, as described by Herlocker et al. [42], *recall* is useless in its pure sense for evaluating *RS*, since it requires knowing all the items that are relevant to a *center user*. The authors of the paper cite previous research by Sarwar et al. [100] that have approximated recall by considering those items held in the test dataset of a cross-validation evaluation as the set of relevant items. They express that this metric might be useful, but should be used carefully. Researchers must be aware of the bias underlying this metric since the items in the test dataset are just a sample of the the items that could be considered relevant. In addition, they point out that this approximated recall should be used in a comparative fashion on the same dataset and not as an absolute measure.

Usually the list of recommended items is ranked from most to less relevant. When that is the case, a useful metric is the Discounted Cumulative Gain [45], which measures how effective the recommendation method is at locating the most relevant items at the top and the less relevant items at the bottom of the recommended list. Discounted Cumulative Gain is defined as

$$DCG = \sum_{i}^{p} \frac{2^{rel_i} - 1}{\log_2(1 + i)} \tag{7.8}$$

Usually *normalized DCG* (*nDCG*) [45] is used more frequently, since it allows one to compare the DCG of lists with different length. It is calculated by normalizing the discounted cumulative gain of an ordered list of recommended items by the ideal order of those items if they were ranked perfectly

$$nDCG = \frac{DCG}{iDCG} \tag{7.9}$$

### 7.4.3   Diversity, Novelty and Coverage

Diversity has been shown to be an important factor in user satisfaction regarding system recommendations [128, 124]. Ziegler et al. study how diversity affects a user's opinion, and they derive the *Intra-list Similarity* metric

$$ILS(P_{w_i}) = \frac{\sum_{b_k \in P_{w_i}} \sum_{b_k \in P_{w_i}, b_k \neq b_c} c_o(b_k, b_c)}{2} \tag{7.10}$$

Higher scores of ILS denote lower diversity. Based on this metric, the authors propose a topic diversification algorithm. The results of offline and a large online user study show that "the user's overall liking of recommendation lists goes beyond accuracy and involves other factors, e.g., the users' perceived list diversity" [128].

On a different approach, Lathia et al. [58] introduced two metrics to measure diversity and novelty respectively. They use these measures to evaluate the *RS* performance when considering the drift in users' preferences over time. The metrics are diversity at depth N (7.11) and novelty (7.12)

$$diversity(L1, L2, N) = \frac{|\frac{L2}{L1}|}{N} \tag{7.11}$$

The ratio $L2/L1$ corresponds to the fraction of elements in the list $L2$ that are not in the list $L1$. The second metric is novelty, which compares the current list $L2$ to the set of all items that have been recommended to date $A_t$

$$novelty(L2, N) = \frac{|\frac{L2}{A_t}|}{N} \tag{7.12}$$

Coverage usually refers to the proportion of items that a *RS* can recommend, a concept also called *catalog coverage*. There are also some alternatives to measure coverage during an off-line or on-line experiment, where it is desirable to weight the items by popularity or utility in order, as described in [106]. The same authors describe coverage from the users' point of view, user coverage, understood as the proportion of users for which the system can produce recommendations, as used by Parra and Brusilovsky in [88].

### 7.4.4 Implicit Feedback and Partial Knowledge of User Preferences

In recent years, the research on *RS* has expanded beyond rating-based systems to cope with systems that do not rely on ratings and, even more, that rely mainly on implicit feedback from the users. Under this scenario, several metrics have been introduced, the most important being the *Mean Percentage Ranking (MPR)*, also known as *Percentile Ranking*. It is used when the knowledge source of user interest is implicit feedback. It is a recall-oriented metric, because the authors that have used it [23] [44] state that precision based metrics are not very appropriate as they require knowing which resources are undesirable to a user. Lower values of *MPR* are more desirable. The expected value of MPR for random predictions is 50%, and thus MPR ¿ 50% indicates an algorithm no better than random.

$$MPR = \frac{\sum_{ui} r_{ui}^t \cdot \overline{rank_{ui}}}{\sum_{ui} r_{ui}^t} \qquad (7.13)$$

Where $r_{ui}$ indicates if the user $u$ consumed the item $i$ and $\overline{rank_{ui}}$ denotes the percentile-ranking of $i$ within an ordered list. In this way, $\overline{rank_{ui}} = 0\%$ means that $i$ is at the top of the list [44].

Another metric intended for implicit feedback datasets is *AP Correlation*. It was introduced by Yilmaz et al. [122] as a modification to Kendall's Tau in order to penalize mistakes made regarding highly relevant items more than for less relevant ones. AP correlation finds the precision between two orders at each index in the list and takes the average of these values

$$\tau_{ap} = \frac{2}{N-1} \cdot [\sum_{i \in I} \frac{C(i)}{index(i) - 1}] - 1 \qquad (7.14)$$

$N$ is the number of ranked items in the list, $C(i)$ is the number of items at an index less than $index(i)$ that are correctly ranked according to the ground truth. *AP correlation* ranges from +1 to -1. One problem with this metric is that it assumes that the ground truth list and the evaluated list give a total order, so when just partial orders are available, it is unusable.

In order to deal with partial orders, the *Expected Discounted Rank Correlation (EDRC)* introduced by Ackerman and Chen [1], combines *AP correlation* with *nDCG* to measure the similarity between two sets of pairwise preferences. Similar to both of them, EDRC emphasizes preserving the order of the user's most preferred items and applying a penalty for less preferred items. This metric tries to solve an important evaluation issue, that has been well introduced but not yet tested.

## 7.4.5   Beyond Metrics: User Study Frameworks

Evaluating the users' experience in *RS* has lagged compared to off-line evaluations, since it has not been standardized and it is usually time-consuming. Only recently, in the Recommender Systems Conference of 2011[8], two user evaluation frameworks were introduced, one by Knijnenburg et al. [48] and the other by Pu et al. [93].

The Knijnenburg et al. framework is characterized by subjective and objective evaluations of the *user experience* (*UX*). Figure 7.1 illustrates the framework. To start the evaluation, they consider *objective system aspects* (*OSA*): algorithms, visual and interaction design of the system, the way recommendations are presented and other traits such as social networking. The *subjective system aspects* (*SSA*) contain the users' perception of the *OSA* which are evaluated with questionnaires: their main objective is showing whether the objective aspects (personalization) are perceived at all.

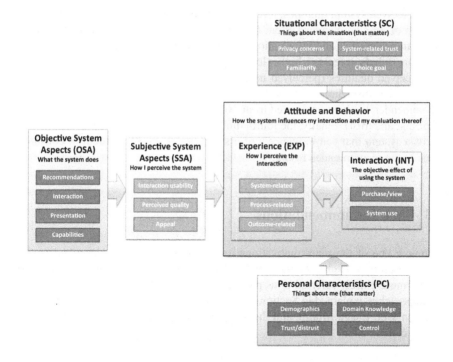

**Fig. 7.1** Kijnenburg's et al. [48] UX evaluation framework

To distinguish between attitude and behavior, Knijnenburg et al. model considers the *experience* (*EXP*) and *interaction* (*INT*). The experience consists of the users' evaluation of the system, also evaluated by questionaries, as *SSA*, and is divided

---

[8] http://recsys.acm.org/2011

into the evaluation of the system, the decision process, and the final decisions made. The interaction is, on the other side, the observable behavior of the user. Finally, the model also considers that experience and interaction are influenced by *personal* (*PC*) and *situational* (*SC*) characteristics. Personal characteristics include demographics, trust, domain knowledge and perceived control. The second set, situational characteristics, depend on the context of the interaction.

In [93], Pu et al. introduced a unifying evaluation framework called ResQue (Recommender systems' Quality of user experience). They built this framework upon well-known usability evaluation models such as TAM (Technology Acceptance Model) and SUMI (Software Usability Measurement Inventory), although Knijnenburg et al. also make use of the first one to develop his framework. Pu et al. cite Kninenburg's framework in their related work but they argue that it fails to relate users perception to the likelihood of user adoption of the system. The main component of ResQue model are four dimensions: the perceived system qualities, users' beliefs, their subjective attitudes, and their behavioral intentions. The first, perceived system qualities, refers to user's perception of the objective characteristics of a recommender system (recommendations quality, interface adequacy, interaction adequacy, and information sufficiency and explicability). The second dimension, Beliefs, refers to a higher level of user perception of the system, influenced by perceived qualities (perceived usefulness, ease of use, and control and transparency). The third dimension, attitudes, refers to the user's overall feeling toward a recommender, likely to be derived from experience (overall satisfaction, confidence inspiring, and trust). Finally, the fourth dimension is about behavioral intentions towards a system that can influence a user's decision to use the system or consume some of the recommended resuts.

## 7.5   Web Recommendations

Although one of the main motivations for developing *RS* is, as described in the abstract of this book chapter, the amount of information available on the Web, Web *RS* are more closely referred to as part of Web Usage Mining in literature than to the approaches explained in Section 7.2. In this section, we aim to provide a bridge between Web Usage Mining and the techniques for building *RS*, i.e., for adaptive web personalization.

### 7.5.1   *Sources of Knowledge for Web Recommendation*

Facca et al. [22] identify three main sources of data for web usage mining: server side, proxy side and client side. At the server level, web server logs are typically found in three ways: Common Log Format, Extended Log Format, or LogML. Other sources from the server side are cookies and TCP/IP packet sniffers. The second main source of data, the proxy side, is similar to the data that can be captured from the server side, but it collects data of groups of users by accessing a large group of serves. Finally, on the client side, Javascript, Java Applets or modified browsers allows us to capture usage data. Some researchers have explored combining other sources of data for web site recommendation, such as Li et al. [63] who combine usage data with content and structure for web site recommendation. More recent research has also shown the use of additional features such as gender, age, and geographical information and they have proved to be beneficial for recommendation, such as Li et al. work on recommending personalized news in the Yahoo! portal [64].

### 7.5.2 Methods for Web Recommendation

In [74], Mobasher identifies the primary methods used in Web *RS* for off-line model building –preferred over memory-based models due to performance and scalability issues–, which are Clustering, Association Rule Discovery, Sequential Pattern Discovery, Markov Models, and Latent Models. Baraglia et al. introduce the *SUGGEST 3.0* system that uses clustering in the first of two steps of their method to produce recommendations [8]. Velasquez et al. also show the effectiveness of clustering for online navigation recommendations [114]. Association rules is a frequently used method in web usage mining and for web recommendations. Mobasher et al. use association rules in conjunction with clustering in [75] to recommend URLs using as dataset the Web site of the Association for Consumer Research logs. Lin et al. make use of association rules with an underlying collaborative approach [66] to produce recommendations. In Markov models, one distinguishing example of *Markov Decision Process* (*MDP*) is the *RS* implemented by Shani et al. in 2005. The authors change the usual approach of seeing the recommendations as a rating prediction problem, and they turn it into a sequential optimization process, implementing it in a commercial system. Regarding latent models, a tensor factorization method for personalized web search recommendation called CubeSVD is introduced by Sun et al. in [109]. An alternative approach is taken by Xu et al., who make use of *Latent Dirichlet Allocation* (*LDA*) in a collaborative Web Recommendation framework to model the latent topic space and discover associations between user sessions via probability inference [119].

### 7.5.3 Evaluation Metrics for Web Recommendation

Evaluation metrics used on Web recommendation do not differ too much from those presented in section 7.4. However, in e-commerce the success of a recommendation method is usually measured by the increase in sales or some signal of user engagement. Mobasher et. al use in [75] precision, coverage, F1 (the harmonic mean between precision and coverage) and weighted average visit percentage (WAVP) to evaluate individual profile effectiveness. This last measure, is defined as:

$$WAVP = \left( \sum_{t \in T_{pr}} \frac{t \cdot pr}{|t|} \right) \left( \sum_{p \in pr} weight(p, pr) \right) \tag{7.15}$$

where $t$ is a specific transaction, $T_{pr}$ the subset of transactions whose elements contain at least one page from $pr$.

## 7.6 Summary

In this chapter, we have presented RS beginning with its historical evolution from the early nineties to present day. In order to give users new to this area an introduction to the most common methods, we provided a classification of the main RS approaches. Then, we focused on the sources of knowledge and evaluation measures used to assess RS performance and quality. In the last section, we tried to bridge the trends seen in RS research with web recommendations, which is the main focus of this book. In the coming years, we expect to see an increasing amount of commercially-available recommender systems, since they are mature in several domains as a technology to engage users and alleviate information overload. New challenges are presented by the growing amount of devices and heterogeneous sources of knowledge available, at different levels of analysis, to provide recommendations. Some of these challenges go beyond the current trends of scalability and big data: data sparsity; how to deal with the new user and new item problems; how to automatically select a recommendation method given an special context; add transparency, diversity and serendipity to RS; how to leverage social networks; how to use implicit feedback; how to assure that off-line evaluation results correlates with on-line user satisfaction; among others. All of these issues remain at least partially unsolved and we expect to see a good deal of applications and research around these topics.

**Acknowledgements.** The first author, Denis Parra, has been partially supported by Conicyt (Comisión Nacional de Investigación Cientíca y Tecnológica, Gobierno de Chile.) to pursue his PhD program in Information Sciences at the University of Pittsburgh.

## References

1. Ackerman, B., Chen, Y.: Evaluating rank accuracy based on incomplete pairwise preferences. In: UCERSTI 2: Workshop on User-Centric Evaluation of Recommender Systems and Their Interfaces, vol. 2 (2011)
2. Adomavicius, G., Sankaranarayanan, R., Sen, S., Tuzhilin, A.: Incorporating contextual information in recommender systems using a multidimensional approach. ACM Trans. Inf. Syst. 23(1), 103–145 (2005)
3. Adomavicius, G., Tuzhilin, A.: Toward the next generation of recommender systems: A survey of the state-of-the-art and possible extensions. IEEE Trans. on Knowl. and Data Eng. 17, 734–749 (2005)

4. Adomavicius, G., Tuzhilin, A.: Context-aware recommender systems. In: Ricci, F., Rokach, L., Shapira, B., Kantor, P.B. (eds.) Recommender Systems Handbook, pp. 217–253. Springer US (2011), doi:10.1007/978-0-387-85820-3

5. Aggarwal, C.C., Wolf, J.L., Yu, P.S.: A new method for similarity indexing of market basket data. In: Proceedings of the 1999 ACM SIGMOD International Conference on Management of Data, SIGMOD 1999, pp. 407–418. ACM, New York (1999)

6. Avesani, P., Massa, P., Tiella, R.: A trust-enhanced recommender system application: Moleskiing. In: SAC 2005: Proceedings of the 2005 ACM Symposium on Applied Computing, pp. 1589–1593. ACM Press (2004)

7. Baltrunas, L., Makcinskas, T., Ricci, F.: Group recommendations with rank aggregation and collaborative filtering. In: Proceedings of the Fourth ACM Conference on Recommender Systems, RecSys 2010, pp. 119–126. ACM, New York (2010)

8. Baraglia, R., Silvestri, F.: An online recommender system for large web sites. In: Proceedings of the 2004 IEEE/WIC/ACM International Conference on Web Intelligence, WI 2004, pp. 199–205. IEEE Computer Society, Washington, DC (2004)

9. Beemanapalli, K., Rangarajan, R., Srivastava, J.: Incorporating Usage Information into Average-Clicks Algorithm, pp. 21–35 (2007)

10. Belkin, N.J., Bruce Croft, W.: Information filtering and information retrieval: two sides of the same coin? Commun. ACM 35(12), 29–38 (1992)

11. Bennett, J., Lanning, S., Netflix, N.: The netflix prize. In: KDD Cup and Workshop in Conjunction with KDD (2007)

12. Berger, H., Denk, M., Dittenbach, M., Pesenhofer, A., Merkl, D.: Photo-Based User Profiling for Tourism Recommender Systems. In: Psaila, G., Wagner, R. (eds.) EC-Web 2007. LNCS, vol. 4655, pp. 46–55. Springer, Heidelberg (2007)

13. Bogers, T.: Recommender Systems for Social Bookmarking. PhD thesis, Tilburg University (December 2009)

14. Bose, A., Beemanapalli, K., Srivastava, J., Sahar, S.: Incorporating Concept Hierarchies into Usage Mining Based Recommendations. In: Nasraoui, O., Spiliopoulou, M., Srivastava, J., Mobasher, B., Masand, B. (eds.) WebKDD 2006. LNCS (LNAI), vol. 4811, pp. 110–126. Springer, Heidelberg (2007)

15. Breese, J.S., Heckerman, D., Kadie, C.: Empirical analysis of predictive algorithms for collaborative filtering. In: 14th Conference on Uncertainty in Artificial Intelligence, pp. 43–52 (1998)

16. Burke, R.: Hybrid recommender systems: Survey and experiments. User Modeling and User-Adapted Interaction 12, 331–370 (2002)

17. Burke, R.D., Hammond, K.J., Young, B.C.: The findme approach to assisted browsing. IEEE Expert. 12, 32–40 (1997)

18. Celma, Í., Serra, X.: Foafing the music: Bridging the semantic gap in music recommendation. Web Semant. 6, 250–256 (2008)

19. Deshpande, M., Karypis, G.: Item-based top-n recommendation algorithms. ACM Trans. Inf. Syst. 22(1), 143–177 (2004)

20. Eirinaki, M., Lampos, H., Vazirgiannis, M., Varlamis, I.: Sewep: Using site semantics and a taxonomy to enhance the web personalization process, pp. 99–108 (2003)

21. Eirinaki, M., Lampos, C., Paulakis, S., Vazirgiannis, M.: Web personalization integrating content semantics and navigational patterns. In: Proceedings of the 6th Annual ACM International Workshop on Web Information and Data Management, WIDM 2004, pp. 72–79. ACM, New York (2004)

22. Facca, F.M., Lanzi, P.L.: Mining interesting knowledge from weblogs: a survey. Data &amp; Knowledge Engineering 53(3), 225–241 (2005)

23. Fang, Y., Si, L.: Matrix co-factorization for recommendation with rich side information and implicit feedback. In: Proceedings of the 2nd International Workshop on Information Heterogeneity and Fusion in Recommender Systems, HetRec 2011, pp. 65–69. ACM, New York (2011)

24. Fazel-Zarandi, M., Devlin, H.J., Huang, Y., Contractor, N.: Expert recommendation based on social drivers, social network analysis, and semantic data representation. In: Proceedings of the 2nd International Workshop on Information Heterogeneity and Fusion in Recommender Systems, HetRec 2011, pp. 41–48. ACM, New York (2011)

25. Fernández-Tobías, I., Cantador, I., Kaminskas, M., Ricci, F.: A generic semantic-based framework for cross-domain recommendation. In: Proceedings of the 2nd International Workshop on Information Heterogeneity and Fusion in Recommender Systems, HetRec 2011, pp. 25–32. ACM, New York (2011)

26. Freyne, J., Berkovsky, S., Smith, G.: Recipe Recommendation: Accuracy and Reasoning. In: Konstan, J.A., Conejo, R., Marzo, J.L., Oliver, N. (eds.) UMAP 2011. LNCS, vol. 6787, pp. 99–110. Springer, Heidelberg (2011)

27. Fry, C., Bow Street Software, Weitzman, L., Ibm: Why surf alone?: Exploring the web with reconnaissance agents, internet (accessed: 20, 2001)

28. Gantner, Z., Rendle, S., Schmidt-Thieme, L.: Factorization models for context-/time-aware movie recommendations. In: Proceedings of the Workshop on Context-Aware Movie Recommendation, CAMRa 2010, pp. 14–19. ACM, New York (2010)

29. Ge, M., Delgado-Battenfeld, C., Jannach, D.: Beyond accuracy: evaluating recommender systems by coverage and serendipity. In: Proceedings of the Fourth ACM Conference on Recommender Systems, RecSys 2010, pp. 257–260. ACM, New York (2010)

30. Golbeck, J.: Trust and nuanced profile similarity in online social networks. ACM Trans. Web 3(4), 12:1–12:33 (2009)

31. Golbeck, J., Hendler, J.: Filmtrust: Movie recommendations using trust in web-based social networks, vol. 6, pp. 497–529. ACM, New York (2006)

32. Goldberg, D., Nichols, D., Oki, B.M., Terry, D.: Using collaborative filtering to weave an information tapestry. Commun. ACM 35, 61–70 (1992)

33. Groh, G.: Recommendations in taste related domains: Collaborative filtering vs. social filtering. In: Proc ACM Group 2007, pp. 127–136 (2007)

34. Guy, I., Carmel, D.: Social recommender systems. In: Proceedings of the 20th International Conference Companion on World Wide Web, WWW 2011, pp. 283–284. ACM, New York (2011)

35. Guy, I., Jacovi, M., Perer, A., Ronen, I., Uziel, E.: Same places, same things, same people?: mining user similarity on social media. In: Proceedings of the 2010 ACM Conference on Computer Supported Cooperative Work, CSCW 2010, pp. 41–50. ACM, New York (2010)

36. Guy, I., Jacovi, M., Shahar, E., Meshulam, N., Soroka, V., Farrell, S.: Harvesting with sonar: the value of aggregating social network information. In: Proceedings of the Twenty-Sixth Annual SIGCHI Conference on Human Factors in Computing Systems, CHI 2008, pp. 1017–1026. ACM, New York (2008)

37. Guy, I., Ronen, I., Wilcox, E.: Do you know?: recommending people to invite into your social network. In: Proceedings of the 14th International Conference on Intelligent User Interfaces, IUI 2009, pp. 77–86. ACM, New York (2009)

38. Guy, I., Zwerdling, N., Carmel, D., Ronen, I., Uziel, E., Yogev, S., Ofek-Koifman, S.: Personalized recommendation of social software items based on social relations. In: Proceedings of the Third ACM Conference on Recommender Systems, RecSys 2009, pp. 53–60. ACM, New York (2009)

39. Hayes, C., Hayes, C., Massa, P., Cunningham, P., Avesani, P., Cunningham, P.: An online evaluation framework for recommender systems. In: Workshop on Personalization and Recommendation in E-Commerce (Malaga). Springer (2002)
40. Herlocker, J., Konstan, J.A., Riedl, J.: An empirical analysis of design choices in neighborhood-based collaborative filtering algorithms. Inf. Retr. 5(4), 287–310 (2002)
41. Herlocker, J.L., Konstan, J.A., Borchers, A., Riedl, J.: An algorithmic framework for performing collaborative filtering. In: Proceedings of the 22nd Annual International ACM SIGIR Conference on Research and Development in Information Retrieval, SIGIR 1999, pp. 230–237. ACM, New York (1999)
42. Herlocker, J.L., Konstan, J.A., Terveen, L.G., Riedl, J.T.: Evaluating collaborative filtering recommender systems. ACM Trans. Inf. Syst. 22, 5–53 (2004)
43. Heymann, P., Koutrika, G., Garcia-Molina, H.: Can social bookmarking improve web search? In: First ACM International Conference on Web Search and Data Mining, WSDM 2008 (February 2008)
44. Hu, Y., Koren, Y., Volinsky, C.: Collaborative filtering for implicit feedback datasets. In: Proceedings of the 2008 Eighth IEEE International Conference on Data Mining, pp. 263–272. IEEE Computer Society, Washington, DC (2008)
45. Järvelin, K., Kekäläinen, J.: Cumulated gain-based evaluation of ir techniques. ACM Trans. Inf. Syst. 20, 422–446 (2002)
46. Jäschke, R., Marinho, L., Hotho, A., Lars, S.-T., Gerd, S.: Tag recommendations in social bookmarking systems. AI Commun. 21, 231–247 (2008)
47. Jawaheer, G., Szomszor, M., Kostkova, P.: Comparison of implicit and explicit feedback from an online music recommendation service. In: HetRec 2010: Proceedings of the 1st International Workshop on Information Heterogeneity and Fusion in Recommender Systems, pp. 47–51. ACM, New York (2010)
48. Knijnenburg, B.P., Willemsen, M.C., Kobsa, A.: A pragmatic procedure to support the user-centric evaluation of recommender systems. In: Proceedings of the Fifth ACM Conference on Recommender Systems, RecSys 2011, pp. 321–324. ACM, New York (2011)
49. Konstan, J.A., Miller, B.N., Maltz, D., Herlocker, J.L., Gordon, L.R., Riedl, J.: Grouplens: applying collaborative filtering to usenet news. Commun. ACM 40(3), 77–87 (1997)
50. Kordumova, S., Kostadinovska, I., Barbieri, M., Pronk, V., Korst, J.: Personalized Implicit Learning in a Music Recommender System. In: De Bra, P., Kobsa, A., Chin, D. (eds.) UMAP 2010. LNCS, vol. 6075, pp. 351–362. Springer, Heidelberg (2010)
51. Koren, Y.: Factorization meets the neighborhood: A multifaceted collaborative filtering model. In: ACM KDD, pp. 426–434 (2008)
52. Koren, Y.: Collaborative filtering with temporal dynamics. In: ACM KDD, Paris, France, pp. 89–97 (2009)
53. Koren, Y., Bell, R., Volinsky, C.: Matrix factorization techniques for recommender systems. Computer 42(8), 30–37 (2009)
54. Koren, Y., Sill, J.: Ordrec: an ordinal model for predicting personalized item rating distributions. In: Proceedings of the Fifth ACM Conference on Recommender Systems, RecSys 2011, pp. 117–124. ACM, New York (2011)
55. Krulwich, B., Burkey, C.: Learning user information interests through extraction of semantically significant phrases. In: Proceedings of the AAAI Spring Symposium on Machine Learning in Information Access, pp. 100–112 (1996)
56. Kuroiwa, T., Bhalla, S.: Book recommendation system for utilisation of library services. Int. J. Comput. Sci. Eng. 5, 207–213 (2010)

57. Lang, K.: Newsweeder: Learning to filter netnews. In: Proceedings of the 12th International Machine Learning Conference, ML 1995 (1995)
58. Lathia, N., Hailes, S., Capra, L., Amatriain, X.: Temporal diversity in recommender systems. In: Proceeding of the 33rd International ACM SIGIR Conference on Research and Development in Information Retrieval, SIGIR 2010, pp. 210–217. ACM, New York (2010)
59. Lee, D.H.: Pittcult: trust-based cultural event recommender. In: Proceedings of the 2008 ACM Conference on Recommender Systems, RecSys 2008, pp. 311–314. ACM, New York (2008)
60. Lee, T., Park, Y., Park, Y.: A time-based approach to effective recommender systems using implicit feedback. Expert Syst. Appl. 34(4), 3055–3062 (2008)
61. Lemire, D., Maclachlan, A.: Slope one predictors for online rating-based collaborative filtering. In: Proceedings of SIAM Data Mining SDM 2005 (2005)
62. Lerman, K.: Social networks and social information filtering on digg. CoRR, abs/cs/0612046 (2006)
63. Li, J., Zaïane, O.R.: Combining Usage, Content, and Structure Data to Improve Web Site Recommendation. In: Bauknecht, K., Bichler, M., Pröll, B. (eds.) EC-Web 2004. LNCS, vol. 3182, pp. 305–315. Springer, Heidelberg (2004)
64. Li, L., Chu, W., Langford, J., Schapire, R.E.: A contextual-bandit approach to personalized news article recommendation. In: Proceedings of the 19th International Conference on World Wide Web, WWW 2010, pp. 661–670. ACM, New York (2010)
65. Lieberman, H.: Letizia: An agent that assists web browsing. In: International Joint Conference on Artificial Intelligence, pp. 924–929 (1995)
66. Lin, W., Alvarez, S.A., Ruiz, C.: Efficient adaptive-support association rule mining for recommender systems. Data Min. Knowl. Discov. 6(1), 83–105 (2002)
67. Manning, C.D., Raghavan, P., Schtze, H.: Introduction to Information Retrieval. Cambridge University Press, New York (2008)
68. Massa, P., Avesani, P.: Trust-aware recommender systems. In: Proceedings of the 2007 ACM Conference on Recommender Systems, RecSys 2007, pp. 17–24. ACM, New York (2007)
69. McDonald, D.W., Ackerman, M.S.: Expertise recommender: a flexible recommendation system and architecture. In: Proceedings of the 2000 ACM Conference on Computer Supported Cooperative Work, CSCW 2000, pp. 231–240. ACM, New York (2000)
70. McNee, S.M., Riedl, J., Konstan, J.A.: Being accurate is not enough: how accuracy metrics have hurt recommender systems. In: CHI 2006 Extended Abstracts on Human Factors in Computing Systems, CHI EA 2006, pp. 1097–1101. ACM, New York (2006)
71. Mladenic, D.: Personal webwatcher: design and implementation (1996)
72. Mladenic, D.: Text-learning and related intelligent agents: A survey. IEEE Intelligent Systems 14(4), 44–54 (1999)
73. Mobasher, B.: Data Mining for Web Personalization. In: Brusilovsky, P., Kobsa, A., Nejdl, W. (eds.) Adaptive Web 2007. LNCS, vol. 4321, pp. 90–135. Springer, Heidelberg (2007)
74. Mobasher, B.: Data Mining for Web Personalization. In: Brusilovsky, P., Kobsa, A., Nejdl, W. (eds.) Adaptive Web 2007. LNCS, vol. 4321, pp. 90–135. Springer, Heidelberg (2007)
75. Mobasher, B., Dai, H., Luo, T., Nakagawa, M.: Discovery and evaluation of aggregate usage profiles for web personalization. Data Min. Knowl. Discov. 6(1), 61–82 (2002)
76. Mobasher, B., Dai, H., Luo, T., Sun, Y., Zhu, J.: Integrating Web Usage and Content Mining for More Effective Personalization. In: Bauknecht, K., Madria, S.K., Pernul, G. (eds.) EC-Web 2000. LNCS, vol. 1875, pp. 165–176. Springer, Heidelberg (2000)

77. Mooney, R.J., Roy, L.: Content-based book recommending using learning for text categorization. In: Proceedings of the Fifth ACM Conference on Digital Libraries, DL 2000, pp. 195–204. ACM, New York (2000)
78. Morita, M., Shinoda, Y.: Information Filtering Based on User Behavior Analysis and Best Match Text Retrieval. In: SIGIR 1994: Proceedings of the 17th Annual International ACM SIGIR Conference, pp. 272–281. Springer-Verlag New York, Inc., New York (1994)
79. Mulvenna, M.D., Anand, S.S., Büchner, A.G.: Personalization on the net using web mining: introduction. Commun. ACM 43, 122–125 (2000)
80. Murakami, T., Mori, K., Orihara, R.: Metrics for Evaluating the Serendipity of Recommendation Lists. In: Satoh, K., Inokuchi, A., Nagao, K., Kawamura, T. (eds.) JSAI 2007. LNCS (LNAI), vol. 4914, pp. 40–46. Springer, Heidelberg (2008)
81. Nakagawa, M., Mobasher, B.: A Hybrid Web Personalization Model Based on Site Connectivity
82. Nasraoui, O., Frigui, H.: Extracting web user profiles using relational competitive fuzzy clustering (2000)
83. Oard, D., Kim, J.: Modeling information content using observable behavior. In: Proc. of the ASIST Annual Meeting, pp. 481–488 (2001)
84. O'Connor, M., Cosley, D., Konstan, J.A., Riedl, J.: Polylens: a recommender system for groups of users. In: Proceedings of the Seventh Conference on European Conference on Computer Supported Cooperative Work, ECSCW 2001, pp. 199–218. Kluwer Academic Publishers, Norwell (2001)
85. Parra, D., Amatriain, X.: Walk the Talk: Analyzing the Relation between Implicit and Explicit Feedback for Preference Elicitation. In: Konstan, J.A., Conejo, R., Marzo, J.L., Oliver, N. (eds.) UMAP 2011. LNCS, vol. 6787, pp. 255–268. Springer, Heidelberg (2011)
86. Parra, D., Brusilovsky, P.: Collaborative filtering for social tagging systems: an experiment with citeulike. In: Proceedings of the Third ACM Conference on Recommender Systems, RecSys 2009, pp. 237–240. ACM, New York (2009)
87. Parra, D., Karatzoglou, A., Amatriain, X.: Implicit Feedback Recommendation via Implicit-to-Explicit Ordinal Logistic Regression Mapping, vol. 1 (2011)
88. Parra-Santander, D., Brusilovsky, P.: Improving collaborative filtering in social tagging systems for the recommendation of scientific articles. In: Proceedings of the 2010 IEEE/WIC/ACM International Conference on Web Intelligence and Intelligent Agent Technology - Volume 01, pp. 136–142. IEEE Computer Society, Washington, DC (2010)
89. Pazzani, M., Billsus, D., Michalski, S., Wnek, J.: Learning and revising user profiles: The identification of interesting web sites. In: Machine Learning, pp. 313–331 (1997)
90. Pazzani, M.J.: A framework for collaborative, content-based and demographic filtering. Artificial Intelligence Review 13, 393–408 (1999)
91. Pollock, S.: A rule-based message filtering system. ACM Trans. Inf. Syst. 6, 232–254 (1988)
92. Preece, J., Shneiderman, B.: The reader-to-leader framework: Motivating technology-mediated social participation. AIS Transactions on Human Computer Interaction 1(1), 13–32 (2009)
93. Pu, P., Chen, L., Hu, R.: A user-centric evaluation framework for recommender systems. In: Proceedings of the Fifth ACM Conference on Recommender Systems, RecSys 2011, pp. 157–164. ACM, New York (2011)
94. Quercia, D., Lathia, N., Calabrese, F., Di Lorenzo, G., Crowcroft, J.: Recommending social events from mobile phone location data. In: Proceedings of IEEE ICDM 2010 (December 2010)

95. Redpath, J., Glass, D.H., McClean, S., Chen, L.: Collaborative Filtering: The Aim of Recommender Systems and the Significance of User Ratings. In: Gurrin, C., He, Y., Kazai, G., Kruschwitz, U., Little, S., Roelleke, T., Rüger, S., van Rijsbergen, K. (eds.) ECIR 2010. LNCS, vol. 5993, pp. 394–406. Springer, Heidelberg (2010)

96. Resnick, P., Iacovou, N., Suchak, M., Bergstrom, P., Riedl, J.: Grouplens: an open architecture for collaborative filtering of netnews. In: Proceedings of the 1994 ACM Conference on Computer Supported Cooperative Work, CSCW 1994, pp. 175–186. ACM, New York (1994)

97. Resnick, P., Varian, H.R.: Recommender systems. Commun. ACM 40, 56–58 (1997)

98. Sahebi, S., Oroumchian, F., Khosravi, R.: An enhanced similarity measure for utilizing site structure in web personalization systems. In: Proceedings of the 2008 IEEE/WIC/ACM International Conference on Web Intelligence and Intelligent Agent Technology - Volume 03, WI-IAT 2008, pp. 82–85. IEEE Computer Society, Washington, DC (2008)

99. Salton, G., McGill, M.J.: Introduction to Modern Information Retrieval. McGraw-Hill, Inc., New York (1986)

100. Sarwar, B., Karypis, G., Konstan, J., Riedl, J.: Analysis of recommendation algorithms for e-commerce. In: Proceedings of the 2nd ACM Conference on Electronic Commerce, EC 2000, pp. 158–167. ACM, New York (2000)

101. Sarwar, B., Karypis, G., Konstan, J., Riedl, J.: Item-based collaborative filtering recommendation algorithms. In: Proceedings of the 10th International Conference on World Wide Web, WWW 2001, pp. 285–295. ACM, New York (2001)

102. Sarwar, B., Karypis, G., Konstan, J., Riedl, J.: Itembased collaborative filtering recommendation algorithms. In: Proc. 10th International Conference on the World Wide Web, pp. 285–295 (2001)

103. Sarwar, B.M., Karypis, G., Konstan, J.A., Riedl, J.T.: Application of dimensionality reduction in recommender system – a case study. In: ACM Webkdd Workshop (2000)

104. Ben Schafer, J., Konstan, J., Riedi, J.: Recommender systems in e-commerce. In: Proceedings of the 1st ACM Conference on Electronic commerce, EC 1999, pp. 158–166. ACM, New York (1999)

105. Senot, C., Kostadinov, D., Bouzid, M., Picault, J., Aghasaryan, A.: Evaluation of group profiling strategies. In: IJCAI, pp. 2728–2733 (2011)

106. Shani, G., Gunawardana, A.: Evaluating recommendation systems. In: Recommender Systems Handbook, pp. 257–297 (2011)

107. Shardanand, U., Maes, P.: Social information filtering: algorithms for automating word of mouth. In: Proceedings of the SIGCHI Conference on Human Factors in Computing Systems, CHI 1995, pp. 210–217. ACM Press/Addison-Wesley Publishing Co., New York (1995)

108. Sinha, R.R., Swearingen, K.: Comparing Recommendations Made by Online Systems and Friends. In: DELOS Workshop: Personalisation and Recommender Systems in Digital Libraries (2001)

109. Sun, J.-T., Zeng, H.-J., Liu, H., Lu, Y., Chen, Z.: Cubesvd: a novel approach to personalized web search. In: Proceedings of the 14th International Conference on World Wide Web, WWW 2005, pp. 382–390. ACM, New York (2005)

110. Takeuchi, Y., Sugimoto, M.: CityVoyager: An Outdoor Recommendation System Based on User Location History. In: Ma, J., Jin, H., Yang, L.T., Tsai, J.J.-P. (eds.) UIC 2006. LNCS, vol. 4159, pp. 625–636. Springer, Heidelberg (2006)

111. Tkalcic, M., Kunaver, M., Kosir, A., Tasic, J.: Addressing the new user problem with a personality based user similarity measure. In: Masthoff, J., Grasso, F., Ham, J. (eds.) UMMS 2011: Workshop on User Models for Motivational Systems: The Affective and the Rational Routes to Persuasion (2011)
112. Tso-Sutter, K.H.L., Marinho, L.B., Schmidt-Thieme, L.: Tag-aware recommender systems by fusion of collaborative filtering algorithms. In: Proceedings of the 2008 ACM Symposium on Applied Computing, SAC 2008, pp. 1995–1999. ACM, New York (2008)
113. Vargas, S., Castells, P.: Rank and relevance in novelty and diversity metrics for recommender systems. In: Proceedings of the Fifth ACM Conference on Recommender Systems, RecSys 2011, pp. 109–116. ACM, New York (2011)
114. Velasquez, J.D., Bassi, A., Yasuda, H., Aoki, T.: Mining web data to create online navigation recommendations. In: Perner, P. (ed.) ICDM 2004. LNCS (LNAI), vol. 3275, pp. 551–554. Springer, Heidelberg (2004)
115. Velásquez, J.D., Palade, V.: Adaptive Web Sites: A Knowledge Extraction from Web Data Approach. IOS Press, Amsterdam (2008)
116. Velsquez, J.D., Palade, V.: Building a knowledge base for implementing a web-based computerized recommendation system. International Journal on Artificial Intelligence Tools 16(05), 793 (2007)
117. Victor, P., De Cock, M., Cornelis, C.: Trust and recommendations. In: Recommender Systems Handbook, pp. 645–675 (2011)
118. Walter, F.E., Battiston, S., Schweitzer, F.: A model of a trust-based recommendation system on a social network. Autonomous Agents and Multi-Agent Systems 16(1), 57–74 (2008)
119. Xu, G., Zhang, Y., Yi, X.: Modelling user behaviour for web recommendation using lda model. In: IEEE/WIC/ACM International Conference on Web Intelligence and Intelligent Agent Technology, WI-IAT 2008, vol. 3, pp. 529–532 (December 2008)
120. Yanbe, Y., Jatowt, A., Nakamura, S., Tanaka, K.: Can social bookmarking enhance search in the web? In: Proceedings of the 7th ACM/IEEE-CS Joint Conference on Digital Libraries, JCDL 2007, pp. 107–116. ACM, New York (2007)
121. Yang, W.-S., Cheng, H.-C., Dia, J.-B.: A location-aware recommender system for mobile shopping environments. Expert Systems with Applications 34(1), 437–445 (2008)
122. Yilmaz, E., Aslam, J.A., Robertson, S.: A new rank correlation coefficient for information retrieval. In: Proceedings of the 31st Annual International ACM SIGIR Conference on Research and Development in Information retrieval, SIGIR 2008, pp. 587–594. ACM, New York (2008)
123. Yu, P.S.: Data mining and personalization technologies. In: Proceedings of the Sixth International Conference on Database Systems for Advanced Applications, DASFAA 1999, pp. 6–13. IEEE Computer Society, Washington, DC (1999)
124. Zhang, Y.C., Séaghdha, D.O., Quercia, D., Jambor, T.: Auralist: introducing serendipity into music recommendation. In: Proceedings of the Fifth ACM International Conference on Web Search and Data Mining, WSDM 2012, pp. 13–22. ACM, New York (2012)
125. Zhou, T., Kuscsik, Z., Liu, J.-G., Medo, M., Wakeling, J.R., Zhang, Y.-C.: Solving the apparent diversity-accuracy dilemma of recommender systems. Proceedings of the National Academy of Sciences 107(10), 4511–4515 (2010)
126. Zhu, T., Greiner, R., HŁubl, G.: An effective complete-web recommender system (2003)
127. Ziegler, C.-N., Golbeck, J.: Investigating interactions of trust and interest similarity. Decis. Support Syst. 43, 460–475 (2007)
128. Ziegler, C.-N., McNee, S.M., Konstan, J.A., Lausen, G.: Improving recommendation lists through topic diversification. In: Proceedings of the 14th International Conference on World Wide Web, WWW 2005, pp. 22–32. ACM, New York (2005)

# Author Index

# Subject Index